Electronic Circuits – Practical Learning

Electronic Circuits

Practical Learning

Nicholas L. Pappas. Ph.D.

A Message about this Text: The subject is essentially endless. The purpose here is to say enough about the subject so that you, the reader, have a running start when you apply this knowledge to your work.

Knowledge of some algebra and elementary calculus would help.[1]

We believe important benefits accrue by doing the experiments carefully, I.e. building the circuits and measuring performance.

A Message from the Author: I have worked continuously in the electronics industry since 1950 except for 11 semesters teaching at San Jose State University (Professor and Chair Computer Engineering 1988-1993). There I discovered my talent for teaching such as it may be. After War2 I attended Lehigh University, and then transferred to Stanford where I earned the MS degree and, while working at HP in the early 1950's, the Ph.D. EE degree. (Somehow I did not get the word and formally apply for the BS degree.) Hardware design has been my principal activity. I learned enough about assembly language, Forth, C and C++ to design the software I needed for my projects. My current activity is designing integrated circuits.

---

[1] Thompson and Gardner, *Calculus Made Easy*, ISBN 0312 185 480 (pbk)
Morris Kline, *Calculus*,   ISBN 0486 404 536 (pbk)
N. L. Pappas, Algebra   ISBN 978 1505 345 704 (pbk)

Preface

If you want to know that a circuit is doing what it is supposed to do, then you need to see the voltage waveforms at each circuit node. That is why an oscilloscope is an indispensable tool that enables practical learning.

Practical learning requires knowing how to use measuring instruments such as an oscilloscope, what measurements to make, how to make the measurements, and how to use the tools of the trade.

This text provides that knowledge, and more, as listed in the table of contents. Acquiring this knowledge involves a significant effort and a reasonable expenditure of cash if you are proceeding on a do-it-yourself basis. Implementing practical learning requires you to spend cash on tools, a breadboard, electrical components, and test equipment.

A problem with lots of so called practical learning is that it does not prepare you to move on to what we consider to be the real stuff – i.e. the theoretical basis of electronics which allows you to tackle almost any electronic design project, which we hope is your desire.

If you simply want to do electronics as an interesting hobby, then this book provides a solid, non trivial, basis for that. Non trivial – that's the key phrase – the essential property of this book.

Assuming you move on to the real stuff [1] [2] [3] the practical learning here is essentially what you would learn on your first design project job. Knowledge that is a practical extension of knowledge learned while acquiring a BS degree in electrical or computer engineering.

There is no question in our mind that if you plowed your way through this book *while pursuing a BSEE* (or equivalent) you would not only do better in the BSEE program you would be ahead of the game while in school and at whatever job you took. In fact you would be in a better position to select a job.

---

[1] N. L. Pappas *Electric Circuits – Analysis and Design* ISBN 978 1494 273 385
[2] N. L. Pappas *Electronic Circuit Design with Bipolar and Mos Transistors* ISBN 978 1495 359 729
[3] N. L. Pappas *Digital Design Logic, Memory, Computers* ISBN 978 1499266764

Electronic Circuits – Practical Learning

Theory teaches you through the experience of others. Theoretical knowledge can often lead to a deeper understand of a concept through seeing it in context of a greater whole and understanding the why behind it. Practical knowledge helps you acquire the specific techniques that become the tools of your trade.

Spice programs are sprinkled throughout the text. Their purpose is to show the correct data you can compare to the data your measurements produce.

Contents

Electronic Circuits – Practical Learning

1 Starting Up

SAFETY FIRST Electricity is silent so be very careful. We remove all metal objects from our hands, wrists, and neck such as rings, watches, and necklaces. If you do not remove them, then know you are taking an unnecessary risk.

Furthermore you do the experiments at your own risk, because there is no way we can supervise your work.

Word to the wise – NEVER GRAB anything, because if its "hot" you will have difficulty letting go.

The AC line voltage is extremely dangerous.

1.1 Items You will Need

Here are most of the items you will need in due course of events.

Spice - Free Spice programs are available on the internet.

Tools - Side cutters and long nose pliers cut and form wire. A wire stripper strips insulation from the ends of wire without damaging the wire. A lead forming tool accurately forms wire leads of parts for insertion into the breadboard holes. Tweezers facilitate picking up and placing small parts. Clip leads connect terminals as needed.
1 reactance chart (to get one do a Google search for "reactance chart")
1 solderless breadboard
1 side cutters, 4"
1 long nose pliers, 4"
1 wire stripper
1 lead forming tool for resistors and other components
1 tweezers, fine point
1 IC DIP extraction tool
? clip leads

Electronic Circuits – Practical Learning

Parts A comprehensive view of what parts are available is found in the Product Index of electronics parts distributors. The electronic circuits experiments only use a small subset of each type of part such as resistors R, capacitors C, inductors L, transformers, and potentiometers.

Associated with every part is a *data sheet*, which presents the part's characteristics in many formats: text descriptions, drawings, maximum ratings, thermal characteristics, tables of electrical characteristics, graphs of typical characteristics, available electrical values, package dimensions, pin assignments, and test circuits.

A *data sheet* in effect tells you what a part is about.

Rarely will you find in a *data sheet* the equivalent circuit of a part given a frequency range, nor any information about the part's parasitic components attached to them.. This information is usually found in *technical articles* or *application notes* issued by the manufacturer.

Every manufacturer has a web site from which you can download data sheets, application notes, white papers and so forth.

Each experiment specifies the parts used in the experiment's circuits.

Equipment What do you need to measure or observe, while evaluating circuit performance? As a minimum you need to measure or observe ohms, DC voltages, DC currents, steady state AC signal voltages, and transient state signal voltages.

You need an oscilloscope so that you can "see" the DC voltages and the signal voltages at circuit nodes. Two channels allow you to compare what is going on at two nodes, such as an input and its corresponding output. For our purposes here a two channel oscilloscope, with 1MHz (mega hertz) bandwidth or better, is satisfactory as well as cost effective.

You need a signal generator, specifically a function generator, to generate the signals driving the circuit under test. Function generator, because you will need sine waves, square waves, triangular waves, and pulses. A function generator with maximum frequency 1MHz or better will do. If more than 1 MHz is available at low cost buy it.

If you buy used equipment make sure it is *refurbished* and in 100% working order

A very important feature of any signal generator is that the signal amplitude does NOT change when frequency is changed. This must be verified when experiments are executed.

You need a DC multimeter, which measures wide ranges of volts, ohms, and amperes. Analog or digital meter? Your choice. We use both types.

To us a Linear Power Supply is a *generator* of zero frequency DC voltages. In this text's experiments you only need ±5V. The minus 5V is needed to implement differential amplifiers.

Emphasis: Do NOT buy a switching power supply.

1.2 The Color Code Identifies Value

An effective way to learn the color code is to sort a pile of resistors by value. *Another way is to use the program colorcode.exe.*

1st band   1st digit
2nd band   2nd digit
3rd band   number of zeros
4th band   tolerance, gold 5%, silver 10%

digit	**0**	**1**	**2**	**3**	**4**	**5**	**6**	**7**	**8**	**9**	
color	*black*	*brown*	*red*	*orange*	*yellow*	*green*	*blue*	*violet*	*gray*	*white*	
10% *values*	100	120	150	180	220	270	330	390	470	560	680	820	1000
5% *values*	100	110	120	130	150	160	180	200	220	240	270	300	330
	360	390	430	470	510	560	620	680	750	820	910	1000	

Examples

22	red, red, black
220,000	red, red, yellow
1,200	brown, red, red
47,000	yellow, violet, orange
910	white, brown, brown
8.2×10^6	gray, red, green

1.3 The Solderless Breadboard and the Power Supply

Solderless breadboards are designed to connect parts together without using solder. A power supply energizes the circuits on the solderless breadboard.

You build a circuit by "plugging in" parts. For example a resistor has 2 leads, which when suitably bent can be inserted into 2 pin holes.

A part's lead (28 to 20 AWG, 0.0126 to 0.0320 inches diameter) is inserted into a hole whose spring loaded metal insert "grabs" the lead. The metal inserts are not visible. *The metal inserts in horizontal rows of 5 holes are shorted together*. Consequently leads inserted in the same 5-hole-row are shorted together. The leads placed in a 5 hole row are the equivalent of leads soldered together.

Figure 101 Part of a Solderless breadboard showing fields of holes.

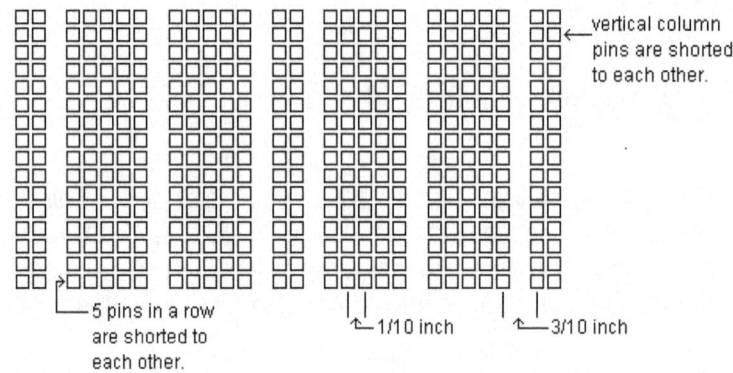

A breadboard is an assembly of *two types of hole patterns* that may be mounted on a metal plate (Figure 101). One type of hole pattern has two 5 × 59 arrays of holes separated by a narrow gutter. The columns of holes parallel and adjacent to the channel are 0.3 inches apart, because 0.3 inches is the IC DIP package minimum pin row spacing. Each array of holes is a column of 59 five hole rows. All holes are on a 0.1 inch grid so that vertical and horizontal separation is 0.1 inch. The five holes in each row are shorted together. The rows are not shorted together.

IC pins are inserted into the board so that the IC straddles the gutter and each IC pin plugs into one hole of one row. Then the four other holes in each "pin" row are available to receive leads, which are automatically connected to the IC pin. In this way a circuit is wired from node to node. Larger ICs have 0.4-inch and 0.6-inch pin row spacing. These are inserted in the same way that the narrower 0.3 inch ICs are inserted; however, the covered up row holes are not available for point-to-point wiring.

Power and Ground The other type of hole pattern is formatted to distribute power and ground. The pattern has two columns of 50 holes. The 50 holes *in each column* are shorted together. The two columns are not shorted together so that one column can distribute 5 volts, for example, and the other column 0 volts (ground). In a large breadboard another pattern across the top of the board has two rows of 40 holes. The 40 holes *in each row* are shorted together.

Above the rows of 40 holes the solderless breadboard with a metal bottom plate has binding posts whose insulated "knobs" unscrew to reveal a hole in the post in which a wire is inserted. The other end of the wire is plugged into a row of 40 holes. In turn jumpers connect the 40 hole rows to the 50 hole columns. (The black post is shorted to the metal base plate, and the red ones are insulated from the base plate.)

Verifying the solderless breadboard "shorts"
Select the multimeter R×10 or higher ohm range, because the R×1 range drains the 1.5 volt battery (use the R×1 range only when you have to).

Connect a pin to each lead of your multimeter. Place one pin in the first hole of a 5 hole row. Place the other pin in each of the other 4 holes in the row. Verify that the resistance is essentially zeros ohms (the short).

Repeat for the columns of holes.

Measure the resistance between rows, which should be infinite (open circuit). Repeat for row to columns, etc. Check out all possibilities to KNOW how the holes are wired.

Electronic Circuits – Practical Learning

Power Supply You can select any *linear ±5V* power supply. We use a linear open frame *±5V* power supply. We have to add a power cord. COVER THE AC TERMINALS WITH TAPE. Plug the power cord into a plug strip outlet. The plug strip on/off switch becomes the power supply on/off switch. Think of a power supply as a constantly recharged battery. *Do not buy a switching supply – it generates too much noise.*

The source impedance R_S of the "battery" is estimated as follows. If a 5V output drops by 2%, or 0.1V, when a 0.25 ampere load current is drawn, then R_S=0.1V/0.25A=0.4 ohms. However if plus is shorted to minus, then potentially I=5/0.4=12.5 amperes, which may or may not flow. This is why unprotected parts are destroyed (and perhaps the power supply).

We have to be very careful to avoid shorting plus to minus.

> *Shut off the power supply OR Disconnect the voltage leads AT THE SUPPLY* when making changes on the solderless breadboard.

Connecting the power supply to the solderless breadboard

Connect the binding posts to the solderless breadboard. The solderless breadboard has "binding" posts whose insulated "knobs" unscrew to reveal a hole in the post. The black post is shorted to the metal base plate, and the red ones are insulated from the base plate. Unscrew the black post and insert a black wire in the post hole. Tighten the black knob to secure the wire. Insert the other end of the wire into a hole in a column of holes that you want to be grounded. Repeat with a red wire from a red post to what becomes the "power supply voltage, the B+" column of holes (e.g. +5V). Repeat for B− (e.g. −5V).

Connect the power supply to the binding post. First turn off the AC power to the supply. Use clip leads or wires to connect the power supply voltage terminals to the solderless breadboard binding posts.

Note: you do not have to use the binding posts.

AFTER you have built a circuit, turn on the AC power to the supply. Turn the power off *before* you make circuit changes.

Use the multimeter to verify the *±5V* voltages.

2 Analog Oscilloscopes

If you want to know that a circuit is doing what it is supposed to do, then you need to see the voltage waveforms at each circuit node.

Electronic equipment can be classified into two categories: analog and digital. Analog equipment works with continuously variable voltages, while digital equipment works with binary numbers that represent voltage samples. Oscilloscopes can be classified similarly – as analog and digital types.

Tektronix produced the first really useful oscilloscopes (after World War 2). To this day most oscilloscopes have an 8cm vertical y axis by a 10 cm horizontal x axis display for the waveforms being measured as well as controls that select the y axis volts/cm scale and x axis time/cm scale.

An oscilloscope consists of four basic systems – the y axis vertical system, the x axis horizontal system, the trigger system, and the probe system. Each system contributes to the oscilloscope's ability to accurately display a reconstructed signal. The front panel of an oscilloscope is divided into four sections: vertical, horizontal, trigger, and display (Figure 201).

Bandwidth Bandwidth is the number one specification emphasized by every oscilloscope manufacturer. Bandwidth defines the frequency range a signal's steady state frequency spectrum must be within for the oscilloscope to be able to accurately display the signal. Oscilloscope bandwidth is defined as the frequency range from 0 to f_{-3dB}. At f_{-3dB}. a steady state sine wave input signal is attenuated by 3 dB (0.707).

Specification A typical oscilloscope specification is shown in a side bar on the next page.

Electronic Circuits – Practical Learning

Figure 201 Two Channel Oscilloscope

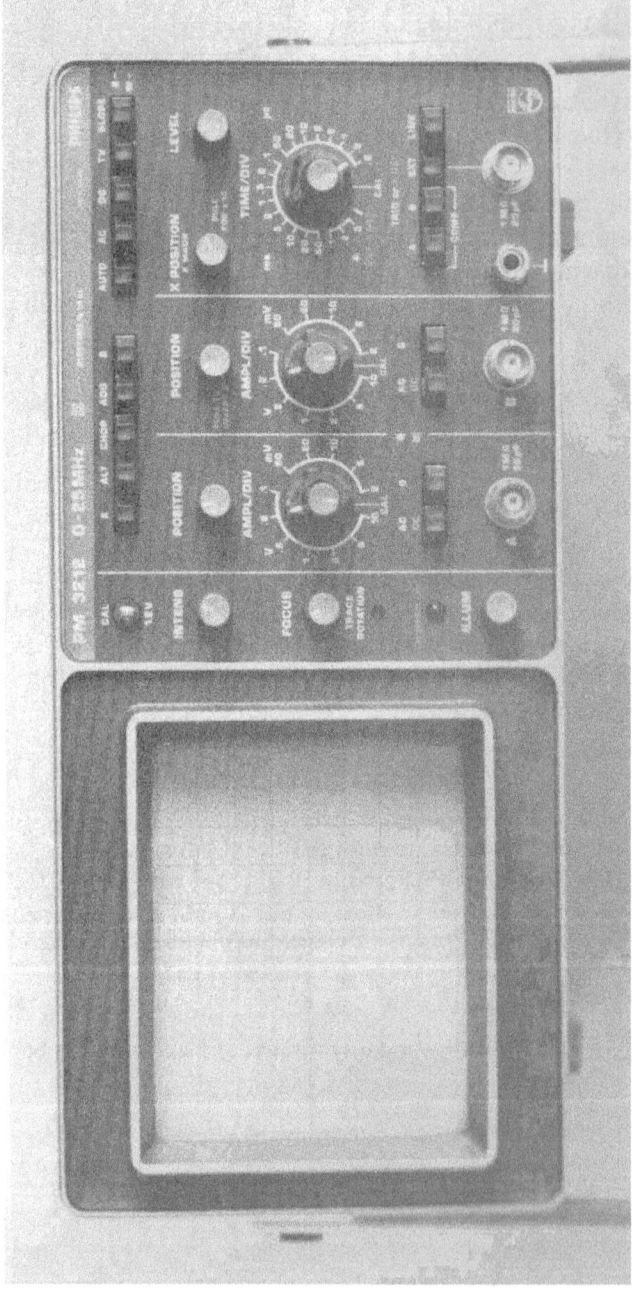

Analog Oscilloscope Specification

The front panel contains the following controls (Figure 201).

Waveform screen display
    Voltage 8 vertical divisions
    Time   10 horizontal divisions.

Vertical controls – identical for 2 channels
    Select AC, DC, Ground – the channel input coupling.
    Select display screen Volts/Division – 2mv to 10v per division
       2, 5, 10, 20, 50 mv per division
       0.1, 0.2, 0.5, 1, 2, 5, 10 volts per division.
    cal – adjust to change Volts/Division scale – not calibrated
    position – sets vertical position of horizontal sweep

Sweep – Seconds/Division – time per division 0.2µs to 0.5s per division
    0.2, 0.5, 1, 2, 5, 10, 20, 50 µs per division
    0.1, 0.2, 0.5, 1, 2, 5, 10, 20, 50 ms per division
    0.1, 0.2, 0.5 s per division
    position – sets start position of horizontal sweep

Trigger controls
    Trigger Source - Channel A, B, external, line
    Trigger Type - Auto. AC, DC, TV, slope (+ edge rise, −edge fall)
    Select - Norm, Auto trigger or View to see the trigger
    Level - the level the signal must cross to acquire a waveform.

2.1 The Vertical System

As a practical matter a two channel oscilloscope displays the waveforms of two signals such as an amplifier's input and output, which can be compared. This is possible, because the channel scales can be selected independently (four and more channel oscilloscopes are available). The vertical scale dimension is volts. Vertical controls are used to select the input coupling, position the waveforms, and scale the waveforms.

Input coupling The input coupling is the connection from a circuit node to an oscilloscope vertical channel input. The coupling can be set to DC, AC, or ground. DC coupling shows all of an input signal. AC coupling blocks the DC component of a signal so that the displayed waveform is centered about zero volts. The AC coupling setting is useful when a signal's DC component is irrelevant. The ground setting disconnects the input signal from the vertical system and grounds the vertical input so that you can see where the zero volts line is located on the screen.

Position control A vertical position control allows you to move the waveform up and down the display to place it where you want it on the screen. In effect this controls the zero volts line.

Volts/Division control The volts-per-division control selects the display y axis scale that varies the size of the waveform on the screen.

For example if, on channel A, the volts/div = 5, then each of the eight vertical divisions represents 5 volts and the entire screen represents 40 volts from bottom to top when the display grid has eight major divisions (Figure 201).

If the setting is 2 millivolts/div on channel B, the entire screen display represents 16 millivolts, and so on.

The maximum voltage you can display on the screen is the maximum available volts/div setting times 8 the number of vertical divisions.

Note: the probe used (Section 2.4), 1X or 10X, also influences the scale factor. You must divide the volts/div scale by the attenuation factor of the probe, which is 1 or 10.

Some oscilloscopes have a fine gain control for scaling a displayed signal to a certain number of divisions.

2.2 The Horizontal System

The horizontal scale x axis dimension is seconds. Horizontal controls select x axis scale of seconds per div, and position the waveform.

Position Control A horizontal position control allows you to move the waveform left and right on the display to exactly where you want it on the screen.

Time/div The time-per-division control selects the display x axis scale that defines the time represented on the screen.

2.3 The Trigger System

Analog oscilloscopes display the signal by painting it on the display after receiving the trigger that starts the horizontal sweep.

An oscilloscope's trigger function synchronizes the horizontal sweep to start at the desired point on the signal waveform. This is how a stable display of the signal is produced.

Trigger controls allow you to stabilize repetitive waveforms and capture single-shot waveforms. The trigger makes repetitive waveforms appear static on the oscilloscope display by repeatedly displaying the same portion of the input signal in the same place on the screen.

The edge triggering, available in analog oscilloscopes, is the basic and most common trigger type. In addition there is threshold triggering offered by some analog oscilloscopes.

Trigger Level and Slope The trigger level and slope controls define the trigger point and determine how a waveform is displayed. The trigger circuit is a comparator circuit. In effect controls select the slope and voltage level on one input of the comparator. When the signal connected to the other comparator input matches the settings, a trigger is generated. The level control determines the signal voltage where the trigger will occur. The slope control determines whether the trigger point is on the rising or the falling edge of a signal. A rising edge has a positive slope and a falling edge hss a negative slope.

Trigger Sources The oscilloscope's horizontal sweep is triggered by an internal source or by an external source. Internal sources are channel A, channel B, and the line voltage.

Trigger Input Coupling AC and DC coupling are provided for.

2.4 The Probes

The input impedance of a vertical channel is usually 1 megohm in parallel with 20 picofarads (20pF). The input impedance of a vertical channel is increased by using a passive probe (Figure 202) that connects a circuit node to the channel input. With the probe the input impedance is 10 megohms in parallel with 11 picofarads (11pF). This probe attenuates the signal by a factor of 10.

Circuit loading by the probe becomes more pronounced at higher frequencies, because the 11pf capacitor impedance ($1/j2\pi fC$) decreases as 1 over frequency. For example the $11\mu\mu F$ (11pF) capacitor impedance magnitude is 1,447 ohms at 10MHz. One needs to calculate actual impedances in order to avoid measurement errors.

Clearly these general-purpose passive probes cannot accurately measure signals with extremely fast rise times. The steady increase in signal clock rates and edge speeds requires higher speed probes with less loading effects. High-speed active and differential probes provide ideal solutions when measuring high-speed and/or differential signals.

Figure 202 Passive probe

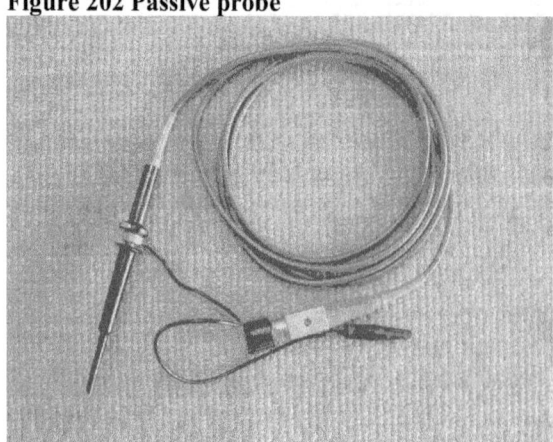

3 Signal Generators

3.1 Waveforms

Periodic and Non-periodic Signals Signals whose waveforms repeat every T seconds are referred to as periodic signals with period T seconds and frequency $f = 1/T$ Hertz (periods per second). Signals that do not repeat are non-periodic signals.

Synchronous and Asynchronous Signals Synchronous signals have a fixed timing relationship. For example clock, data and address signals inside a computer are synchronous signals. Asynchronous signals are signals between which no timing relationship exists. For example, signals produced by pressing a keyboard key.

Sine Waves The sine wave is the fundamental wave shape, because the derivative of a sine wave is a cosine wave, the derivative of a cosine wave is a sine wave, and the sine and cosine waveforms are identical. When the input signals are sine waves the transients produced by the derivatives have the same waveform. Consequently the transients do not disrupt the circuit's steady state behavior.

Periodic sine waves are produced by many sources. The electric utility power system in most countries is based on 50 or 60 cycle per second sine waves. That is why the voltage at any wall outlet is a sine wave. Signal generators produce sine waves as well as other waveforms.

Square and Rectangular Waves The square wave signal is a periodic waveform that has two voltage levels 0 and V_{max}, or $-V_{max}$ and V_{max}. The square wave is at either voltage half the time. The rectangular wave is a modified square wave whose 0 and V_{max}, or $-V_{max}$ and V_{max}, time intervals have different time durations.

Sawtooth Waves Periodic sawtooth waves are produced by non-linear circuits such as the horizontal sweep of an analog oscilloscope. The linear transition of the "saw" voltage level is followed by a step down to 0 volts. The transitions are referred to as ramps and steps.

Electronic Circuits – Practical Learning

Step Functions Non periodical step function signals are referred to as one shot transient signals. A step is a sudden change in voltage, similar to the voltage change you would see if you closed a power switch.

Pulse Trains A pulse train is a periodic sequence of pulses. A pulse whose width is τ seconds is created by a positive step function followed by a negative step function τ seconds later. A pulse produces sudden changes in voltage similar to the voltage changes created by closing and opening a power switch.

Triangular wave This is a periodic waveform not used in any circuit we know of. The positive going linear transition of a "saw" voltage level is followed by a negative going linear transition of a "saw" voltage level.

Figure 301 Function Generator also known as a Signal Generator

3.2 Selecting Signal Parameters

The function generator waveform selection is sine, square, or triangular (Figure 301). The dial varies frequency from 0.2 to 2 times the selected range. The output is TTL format or a 50 ohm source. Adjustable waveform properties are amplitude, voltage offset, symmetry, and frequency sweep over a range.

4 Power Supplies

4.1 Batteries

Standard 1.5 volt and 9 volt low power batteries are available. Batteries may be connected in series for increased power supply voltage. A battery stores charge, which is quoted as so many milliampere–hours (maH) capacity such as 3000mAH when discharge current is 25mA for battery size AA. Data sheets are available on the internet (Figure 401).

Battery voltage decreases when battery charge flows into a circuit. Consequently the battery powered circuit design must accommodate a range of power supply voltage. Battery source impedance is of the order of 300 milliohms.

4.2 Switching Supplies

A wide variety of regulated switching power supplies are available with output voltages ±5v, ±12v, ±15v, ±24v, etc in various combinations. Available power ranges from 20 watts to 3000 watts. Given voltages and powers Ohm's law produces the currents.

There is a catch. The DC voltage output also contains a small AC voltage at the switching frequency, which creates problems when high gain amplifiers are involved unless special filtering measures are taken.

4.3 Linear Supplies

A wide variety of regulated and unregulated linear power supplies are available with output voltages ±5v, ±12v, ±15v, ±24v, etc in various combinations with whatever power is needed.

Electronic Circuits – Practical Learning

Figure 401 AA Battery Data Sheet

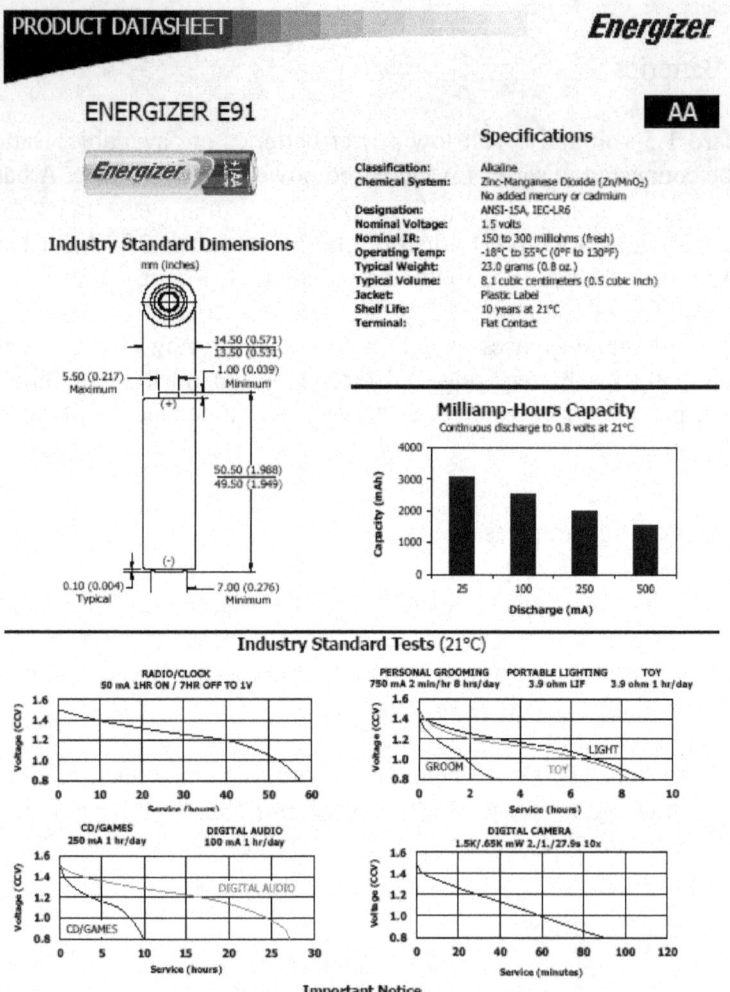

5 When is an Electric Circuit a Distributed Circuit?

There are two basic classes of circuit laws (1) connection constraints, which are Kirchhoff's' laws showing how currents and voltages in any circuit relate to each other and (2) voltage-current *vi* constraints for resistors, capacitors, inductors, transistors, and any other components showing how current relates to voltage in each component.

There are two general analysis methods, node and mesh, for currents and voltages that are assembled from resistors R, capacitors C, inductors L, mutual inductance M, and dependent sources of voltage and current created by transistors for example.

These methods assume that voltage and current changes in a circuit take place throughout the circuit *at the same time* (instantaneously) The world says the circuit is lumped, and the methods are the simplified *lumped methods.*[1]

However any change in a voltage has to *propagate* throughout the circuit. Consequently the *same time* assumption is required in order to use lumped methods. Required, because in the real world propagation of electric signals occurs at the *finite* speed of light (30cm/nanosecond = 30cm/ns = 11.81inches/ns in free space).

Signal propagation takes place in any circuit whenever a voltage at any node changes. The voltage change can take many forms such as a sine wave or step function.

To examine the consequences of the 30cm/ns free space finite speed let a signal generator be located at the left edge of a 15cm long printed circuit board.

Step Function The leading edge of a step function emitted by the generator will travel 15cm in 1ns on printed circuit board traces. The speed is reduced by $1/\sqrt{\varepsilon}$ the square root of the board's dielectric constant ($\varepsilon = 4$). If the leading edge rise time is also 1ns, then 1ns later the leading edge of the voltage waveform spans the printed circuit board (Figure 501).

---

[1] Nicholas L Pappas, *Electric Circuits – Analysis and Design* ISBN 1494273381

Electronic Circuits – Practical Learning

To understand this consider the following. As the signal generator output starts a step function of voltage at the left edge, the voltage at the left edge rises from zero to maximum value in 1ns, while the zero value of the step has traveled 15cm. Consequently the various voltages of the step function are distributed across the 15 cm length of the printed circuit board as shown in Figure 501. This means the 1ns rise time requires circuit analysis as a *distributed* circuit.

(1a) Length L of a step function on a pc board $= \dfrac{\text{rise time } T_r \text{ ns}}{\text{travel time delay ns/cm}} = T_r \times \text{speed}$

(1b) For example $L = \dfrac{1 \text{ns}}{1 \text{ns}/15 \text{cm}} = 1 \text{ns} \times \dfrac{15 \text{cm}}{1 \text{ns}} = 15 \text{cm}$

Figure 501 Step Function Voltage across a 15cm circuit board 1ns later

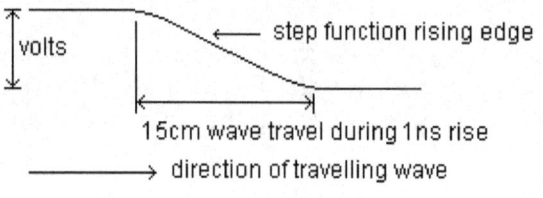

However, if the rise time is 100ns, then the step function "length" is 1500cm not 15cm. Consequently the leading edge voltage across the 15cm board varies by only 1% (1% of 1500=15) anytime during the rise time so that there is essentially the same voltage across the board at *any time* during the rise. Then the circuit can be considered as *lumped*, and the simplified lumped methods can be used.

(2a) *If a signal is periodic, then the signal frequency* $f = \dfrac{1}{\text{period } T}$

(2b) $T = 1 ns \quad \rightarrow \quad f = \dfrac{1}{T} = \dfrac{1}{10^{-9}} = 10^9 = 10^3 \times 10^6 = 1000 \, MHz$

(2c) $T = 100 ns \quad \rightarrow \quad f = \dfrac{1}{T} = \dfrac{1}{100 \times 10^{-9}} = 10^7 = 10^1 \times 10^6 = 10 \, MHz$

Sine Wave A sine wave with 1ns period will span 15cm across the 15 cm circuit board (Figure 502). Whereas only 1% of a sine wave with 100ns period will span 15cm across the 15 cm circuit board at any time. That 1% will be the same voltage across the board within 1%. The 1000MHz sine wave analysis requires the distributed method, and the 10MHz sine wave analysis can use the lumped method.

5 When is an Electric Circuit a Distributed Circuit?

Figure 502 Sine Wave Voltage across a 15cm circuit board (f=10⁹, T=1ns)

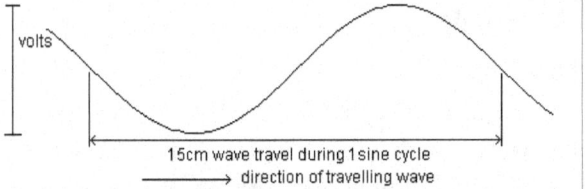

15cm wave travel during 1 sine cycle
⟶ direction of travelling wave

> In other words, given the size of the circuit, the method one can use depends on the rise time of a step function or the frequency of the source or the time taken by any change.

Assume circuit dimensions are 20cm. A voltage rising in 1000ps (1ns) travels about 15cm on a printed circuit board. Therefore at any specific time during the rise the voltage is different at different parts of this 20cm *distributed* circuit. However, if the circuit dimensions are 1cm, then the voltage at any instant of time is essentially the same throughout the 1cm circuit. Now this is a lumped circuit where, for example, a transmission line would be just two zero ohm wires.

The wavelength λ is the distance the periodic electromagnetic wave travels at the velocity of light c in the medium in one period T = 1/f.

Examples: Consider the power cord on any electrical device. A typical power cord is a transmission line that has two parallel wires that connect a 60 cycle AC power source to a device. The 2 meter long cord is NOT a distributed circuit, because the 60 cycle wavelength is 5 million meters.

(3a) $wavelength\, \lambda = velocity\, of\, light\ c \times time\ \dfrac{1}{f} = 3 \times 10^8 \dfrac{meters}{second} \times \dfrac{1}{60}\ seconds$

(3b) $\lambda = 0.05 \times 10^8 = 5 \times 10^6\ meters$

Examples - The wave length on FR-4, a printed circuit board substrate, at 1GHz using an effective dielectric constant of 4 is

(4) $\lambda = \dfrac{c}{f\sqrt{\varepsilon_{eff}}} = \dfrac{3 \cdot 10^{10}\, cm}{1 \cdot 10^9 \sqrt{4}} = \dfrac{30}{2} = 15cm$

And, the wavelength at 10 GHz with a dielectric constant of 10 is

(5) $\lambda = \dfrac{c}{f\sqrt{\varepsilon_{eff}}} = \dfrac{3 \cdot 10^{10}\, cm}{10 \cdot 10^9 \sqrt{10}} = \dfrac{3}{\sqrt{10}} = 0.95cm$

6 Electric Circuit Keywords

6.1 Power

The unit of power P is the watt, which is measured in Joules per second (J/s) where a Joule is the standard unit of energy. We will show later that the power P dissipated by a resistor is the product of voltage and current. $P = vi$ when v and i are "in phase." This means the power in watts it dissipates is expressed as $P = i^2R$ or $P = v^2/R$. These expressions are derived from Ohm's Law where $v=iR$.

$$P = i \times v = i \times iR = i^2R \quad \text{or} \quad P = i \times v = \frac{v}{R} \times v = \frac{v^2}{R}$$

6.2 Electronic Signals

Electronic signals are represented either by voltage or current. The time dependent characteristics of voltage or current signals can take many forms such as constant voltage or constant current (DC), sinusoidal (AC), square wave, linear ramps, and pulse width modulated signals.

Sinusoidal signals are perhaps the most important signal forms since once the circuit response to sinusoidal signals are known, the result can be generalized to predict how the circuit will respond to a much greater variety of signals using the Laplace transform mathematical tool (Appendix A2).

A sinusoidal signal is specified by its amplitude (A), angular frequency (ω), and phase (φ) as $v(t) = A \sin (\omega t + \varphi)$. An important property of sinusoids is that their derivative is also a sinusoid.

When working with sinusoidal signals, the mathematical manipulations often involve computing the effects of the circuit on the amplitude and phase of a signal, which has the form sin ωt or cos ωt. Operations involving these sinusoidal functions are simplified when the complex domain mathematical construct is used (Appendix A1). The sinusoidal signal from the above equation, when expressed in the complex domain, becomes the complex exponential $v(t) = A(sin\ \omega t + jcos\ \omega t) = Ae^{j\omega t}$. The amplitude and phase of the signal are described by the complex constant A, where $A = A_o\ e^{j\varphi}$. As will become clear the complex representation of electronic signals greatly simplifies the analysis of electronic circuits.

6.3 Impedance

The general term for the ratio of voltage to current is impedance Z where *v*=*Zi*. In a circuit assembled only with resistors all of the voltages *v*, currents *i* and resistors R are real numbers, because Ohm's Law (*v* = *iR*) that relates these three types of variables consists of real numbers. The variables are "in phase." Consequently the impedance Z of resistors is R.

What are the impedances, the ratios of voltage to current. of inductors and capacitors? We could find the answers via the Laplace Transform $\mathcal{L}$. However we are not ready for the Laplace Transform (Appendix A2). We are ready for the unorthodox Heaviside who showed that variable *t* for time is replaced by complex frequency variable *p* when *p* is the differential operator *d/dt*.

A complex frequency variable such as p has real and imaginary parts: p = σ+jω, where j is referred to as the imaginary unit (Appendix A1). It is natural to associate σ+jω with the point whose rectangular Cartesian coordinates are (σ, ω) in the σ,ω plane where σ,ω replace the usual x,y coordinates. When σ=0, p=jω and ω is defined in p=jω=j2πf where f is the real frequency as in sin ωt = sin 2πft.

Clearly v/i ratios for L and C in the time domain are not possible because derivatives are involved. However, v/i ratios have meaning in the p complex frequency domain. The RLC voltage-current constraints are

$$\boxed{v(t) = Ri(t) \text{ (page 26)}} \quad \boxed{v(t) = L\frac{di(t)}{dt} \text{ (page 39)}} \quad \boxed{i(t) = C\frac{dv(t)}{dt} \text{ (page 33)}}$$

and impedances are calculated as follows.

$$v(t) = Ri(t) \rightarrow v(p) = Ri(p) \quad \text{so that } R = \frac{v(p)}{i(p)} = z_R$$

$$v(t) = L\frac{di(t)}{dt} \rightarrow v(t) = Lp\,i(t) \Rightarrow v(p) = Lp\,i(p) \rightarrow pL = \frac{v(p)}{i(p)} = z_L$$

$$i(t) = C\frac{dv(t)}{dt} \rightarrow i(t) = Cp\,v(t) \Rightarrow i(p) = pC\,v(p) \rightarrow pC = \frac{i(p)}{v(p)} = \frac{1}{z_C}$$

The generic name for any V(p)/I(p) ratio is impedance z, and the dimension of any v/i ratio is ohms. The generic name for the reciprocal ratio i/v is admittance y, and its dimension is Siemens (it used to be mhos, which we prefer and use). By definition y = 1/z.

7 Passive Components

We start with the resistor R, capacitor C, inductor L, and transformer passive components that appear in almost all circuits. Passive components do not contain energy sources or dependent energy sources.

7.1 Resistor

A resistor R is a passive two-terminal component representing an energy sink. The value of a linear resistor R is invariant with time, and not influenced by the current flowing through it, or the voltage difference between its two terminals. The jargon for voltage difference is "the voltage drop across the resistor." An irreversible conversion of electric energy into heat takes place in a resistor. That is why resistors are designed to dissipate a maximum i^2R power such as a ½ watt, which means do not exceed the ½ watt rating to avoid burning up the resistor. The SI unit of resistance is the ohm. This unit is named after Georg Simon Ohm (1787-1854).

Resistors are used in almost every circuit. Resistors have a linear voltage-current vi relationship obeying Ohm's law where voltage $v = iR$. The vi relationship defines the current flow i produced by an applied voltage v.

The unit of resistance R is the *ohm*, which is represented by the capital letter omega (Ω). Standard resistor values range from 1 Ω to 22 MΩ in ±1%, ±5%, and ±10% tolerances. The voltage v across the resistor R (the voltage drop) equals iR. The power P dissipated in R equals vi (see sidebars on pages 23 and 24).

In practice values of resistance range from about 10^{-3} ohms (1 milliohm) to 10^9 ohms (1000 megohms).The SI unit of conductance is the Siemens that replaced the mho (ohm spelled backwards). We prefer to use mho.

This is about analysis of circuits with resistors. We show that resistors (in fact any combination of components) can be wired in series, in parallel, and in series-parallel. Then we show to how to find solutions by measurements of voltage v and current i in these circuits.

Voltage and Current

Voltage Voltage is the difference in electrical potential between two points in space. It is a measure of the amount of energy gained or lost by moving a unit of positive charge from one point to another.

It is important to keep in mind that voltage is not an absolute quantity; it is the *difference* in value of the potential energy (divided by the charge) at two points in space. I.e. voltage units are Joules per Coulomb.

In a lumped electrical or electronic circuit, the electromagnetic problem of voltages at arbitrary points in space is reduced to voltages between nodes of circuit components such as resistors, capacitors, and transistors.

Voltage is energy/charge. Dimensional analysis confirms this.

$$voltage \; v = electric \; field \; E \times distance \; d$$

$$v = E \times d = \frac{F \times d}{q} = \frac{newton \times meter}{columb} = \frac{joules / meter \times meter}{coulomb} = \frac{joule}{coulomb}$$

> *Definition: A voltage difference of one volt exists between two nodes of a circuit when one joule of energy is required to move one coulomb of charge through the circuit from one node to the other node.*

$$v = \frac{dj}{dq} \qquad units: 1\,volt = 1\frac{joule}{coulomb} \qquad i.e \quad 1V = 1\frac{J}{C}$$

Current Electric currents are electrons in motion with a net flow in some direction. Electric current is the charge passing through a cross section in some unit of time. Current i delivers charge dq in time dt when it flows through a wire or circuit component.

$$i = \frac{dq}{dt} \qquad units: 1\,Ampere = 1\frac{Coulomb \; of \; charge}{second} \qquad i.e. \quad 1A = 1\frac{C}{s}$$

Voltage and Current Sources

There are two kinds of energy sources in electronic circuits: voltage sources and current sources. When connected to an electronic circuit, an *ideal* voltage source maintains a given voltage between its two terminals by providing any amount of current required by the circuit. Similarly, an *ideal* current source maintains a given current to a circuit by providing any amount of voltage across its terminals required by the circuit.

Voltage and current sources are independent or dependent. Their respective circuit symbols are shown in Figure 701. Independent sources are usually shown as a circle while dependent sources are usually shown as a diamond. Independent sources can have a DC output or a functional output such as a sine wave, a square wave, an impulse, or a linear ramp. Dependent sources can be used to implement a voltage or current which is a function of some other voltage or current in the circuit. Dependent sources are often used to model active circuits such as transistors.

Figure 701 Independent and Dependent Sources

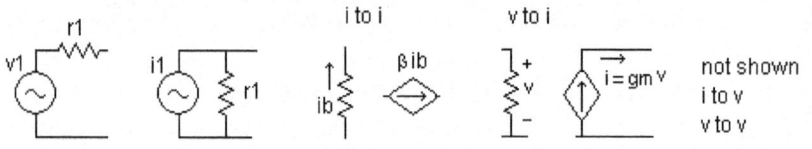

Resistance When a voltage is applied across a conductor, a current will flow. The ratio of voltage v to current i is referred to as resistance R. For most metallic conductors, such as copper, the relationship between voltage v and current i is linear. Stated mathematically, this relationship, or property, is Ohm's law where $R=v/i$. Some electronic components, such diodes and transistors, do not obey the linear Ohm's law, because they have a non-linear current-voltage relationship.

Ground An often used term in circuits is the word *ground*. **Figure 702** The ground is a circuit node to which all voltages in a circuit are referenced. In a constant voltage supply circuit, one terminal from each voltage supply is typically connected to ground, or is *grounded*. For example, the negative terminal of a positive power supply is usually connected to ground so that any current drawn out of the positive terminal can be put back into the negative terminal via ground. Ground is 0 volts. Figure 702 shows the circuit symbol used for ground. The node number 3 is arbitrary.

Oscilloscope DC Voltage Measurements

Voltage is the electric potential, expressed in volts, between two nodes in a circuit. Usually one of these nodes is ground (zero volts).

One basic way to measure a voltage is to count the number of divisions a waveform spans on the oscilloscope's vertical scale.

DC Voltage DC voltage is jargon for a constant voltage such as ground (0 volts) or a 5 volt power supply voltage.

Connect the Channel 1 probe to the node whose DC voltage is to be measured from the node to ground. To simplify the discussion let the DC voltage be 5 volts.

Select GND (0 volts) as the input coupling. Adjust the position control so that the horizontal line on the screen is across the bottom of the screen. The line now represents 0 volts.

Select DC as the input coupling. Adjust the channel *volts/div* control so that the DC signal is visible and spans several divisions. This will be the 1v/div setting since we know we are measuring 5v. The DC signal should be a line across the screen 5 divisions above the 0 volt line at the screen bottom.

For solutions to "resistor" problems apply
1) The concepts of current, voltage and voltage drop
2) Kirchhoff's Laws – the connection constraints
3) Ohm's Law – v = iR the resistor vi constraint.

Electronic Circuits – Practical Learning

7.1.1 *Resistors in Series*

The purpose here is to show that input voltage is divided (allocated) by the resistors, and to emphasize the fact the same current flows in components that are in series.

Theory Kirchhoff's voltage law KVL

$$0 = -v_S + v_1 + v_2 + v_3 \implies v_S = v_1 + v_2 + v_3$$

where the sum of the voltages around a closed loop is zero. The polarity of source voltage v_S is the opposite of the v_B voltages. This is why v_S has a minus sign (Figure 703) and equals the sum of the branch voltages v_B.

Resistor vi constraint - Ohm's Law $v = iR$

Resistors in series add - $R_{series} = R_1 + R_2 + R_3$ because

$$v_S = v_1 + v_2 + v_3 = i_s R_1 + i_s R_2 + i_s R_3 = i_s (R_1 + R_2 + R_3) = i_s (R_{series})$$

Practice: Construct the 703 circuit in Figure 7031. Let V_s=5V, R_1=1K, R_2=2.2K. Use a multimeter or an oscilloscope (page 25). In Figure 7031 measure vb1 (1.56V), vb2 (3.44V). Calculate the current (i =v/R). Measure the current (1.56mA). Reminder: when evaluating results take into account the fact that the resistors have ±x% tolerance.

Construct the 704 circuit in Figure 7041. In Figure 7041 measure vb₁ (0.88V), vb₂ (1.29V), vb₃ (2.84V). Calculate the current. Measure the current (1.29mA).

Figure 703

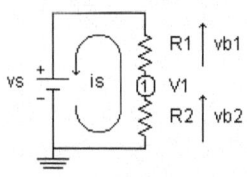

Figure 704

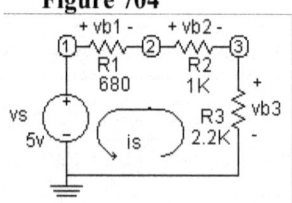

Figure 7031 Circuit for Figure 703

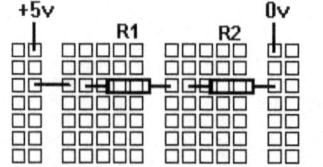

Figure 7041 Circuit for Figure 704

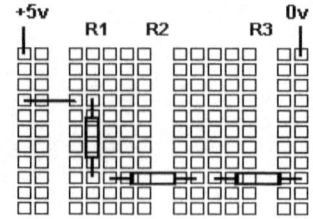

7.1.2 *Resistors in Parallel*

The purpose here is to show that current in any parallel component is independent of current in the other parallel components.

Theory Kirchhoff's current law KCL

$$0 = -i_S + i_1 + i_2 + i_3 \implies i_S = i_1 + i_2 + i_3 \quad \text{(Figure 705)}$$

Resistor vi constraint - Ohm's Law $\quad v = iR$

Using the vi constraint

$$i_S = \frac{v}{R_1} + \frac{v}{R_2} + \frac{v}{R_3} = v\left(\frac{1}{R_1} + \frac{1}{R_2} + \frac{1}{R_3}\right)$$

Resistors in parallel

$$\frac{1}{R_{parallel}} = \frac{1}{R_1} + \frac{1}{R_2} + \frac{1}{R_3}$$

Two Resistors in parallel

$$\frac{1}{R_p} = \frac{1}{R_1} + \frac{1}{R_2} = \frac{R_1 + R_2}{R_1 R_2}$$

Solving for R$_2$

$$R_p = \frac{R_1 R_2}{R_1 + R_2} \implies R_2 = \frac{R_1 R_p}{R_1 - R_p}$$

Example: create a 3.1 non standard resistor value (page 3). Start with 6.8.

$$R_2 = \frac{R_1 R_p}{R_1 - R_p} = \frac{6.8 \times 3.1}{6.8 - 3.1} = \frac{6.8 \times 3.1}{3.7} = 5.7 \ \ use \ 5.6 \ \ then \ R_p = \frac{6.8 \times 5.6}{6.8 + 5.6} = 3.07$$

Practice: Construct the circuit Figure 7051 shown in Figure 705. Let V$_s$=5V, R$_1$=R$_2$=R$_3$=10K. Add one resistor at a time and use a multimeter to measure the current (0.5mA, 1mA, 1.5mA). Each resistor draws 0.5mA - right?

Figure 705   **Figure 7051 Circuit for Figure 705**

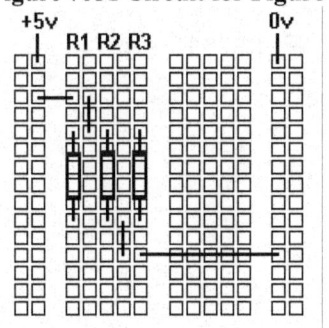

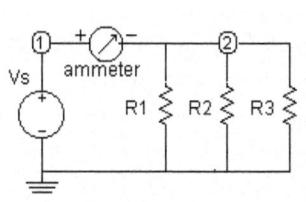

Electronic Circuits – Practical Learning

7.1.3 *Resistors in Series-Parallel Combinations*

The purpose here is to assemble, and evaluate a series parallel circuit that is a 2 mesh circuit..

Practice: Select R_1 = 2.2KΩ, R_2 = 1KΩ, R_3 = 2.2KΩ, and R_4 = 10KΩ (Figure 706). Use a multimeter to measure the actual R values.

Figure 706 Two Mesh Circuit

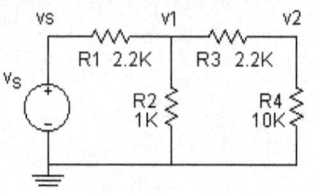

Assemble the four resistors as shown in Figure 7061, and let V_S be a 5v power supply. Measure and record the power supply voltage. Measure the voltages across each resistor (V_{R1}=3.52V. V_{R2}=1.48V, V_{R3}=0.27V, V_{R4}=1.21V)

A formal analysis produces input to output *transfer function* v_2/v_S that shows v_2/v_S = 1.212/5 for the values in Figure 706. Compare to measured v_2/v_S.

Figure 7061 Circuit for Figure 207

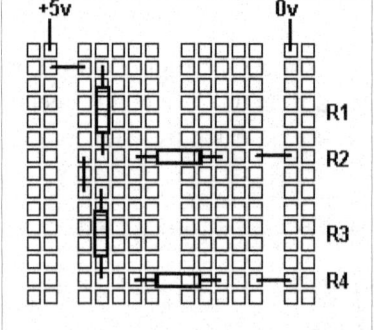

28

7.1.4 *Potentiometers (a.k.a. "pots")*

The purpose here is to show that a potentiometer, a "pot", is a variable divider of voltage.

Practice: Construct the 707 circuit in Figure 7071. Let R=1K. Connect the multimeter to measure V_1. Rotate the 1K pot arm full ccw. Measure V_1 (0V). Rotate pot arm cw slowly, and watch voltmeter reading increase from 0V to almost 2.5V. Why almost?

Change the circuit to Figures 708, 7081. Rotate the 1K pot arm full ccw. Measure V1 (0V). Now rotate the pot arm full cw. Measure V1 (5V). What changed?

A common application of a pot is to vary the gain, transmission, of a voltage amplifier. In such a case the 5v is replaced by a signal voltage.

Figure 707	Figure 708	Figure 709

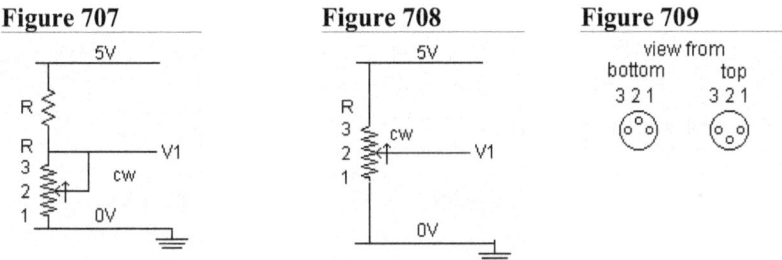

Figure 7071 Circuit for Figure 707	Figure 7081 Circuit for Figure 708

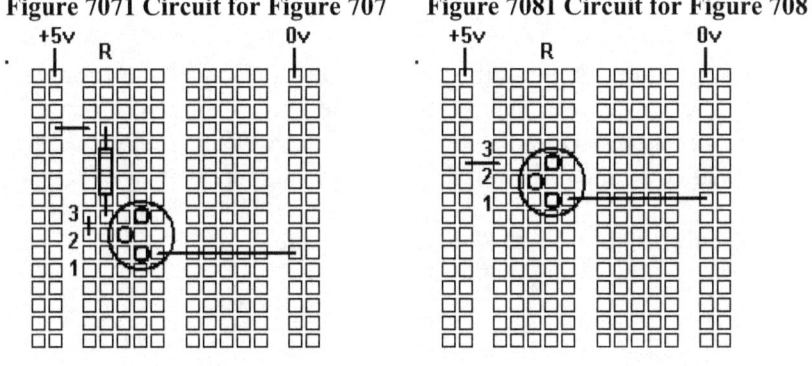

Electronic Circuits – Practical Learning

7.1.5 *Multimeter Voltage Input Impedance*

The purpose here is to emphasize the fact every instrument has an input impedance that becomes part of the circuit. Assemble circuit Figure 7101 on the breadboard

Practice A Simpson multimeter has 20K input resistance per scale volt. Select the 10v scale so that the multimeter input resistance is 200K. Connect the multimeter to nodes 2 and ground. Measure voltages to show that V_2=3.33V, V_1=5V. The 100K and 200K resistors "divide" the 5V into 3 parts: 1 part (1.67V) across the 100K and 2 parts (3.3V) across the 200K (the multimeter).

Figure 710 Resistor   **Figure 7101 Circuit for Figure 710**

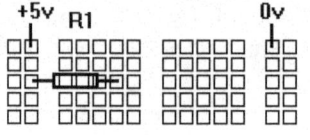

Note: If your multimeter is different, then change the above to suit.

7.1.6 *Multimeter Current Input Impedance*

The purpose here is to know that an ammeter has a very low impedance. In fact the ammeter has about 50mV across its terminals when current has full scale value. You decide if 50mV is negligible in a circuit.

Practice Wire the circuit Figure 7111 on the breadboard *before* connecting the +5V power supply. Select the 1mA scale. Now connect the +5V power supply. Does the meter read 0.5mA? Why? Show that the power consumed by the resistor is 2.5mW. Calculate it two ways.

Figure 711 Current   **Figure 7111 Circuit**

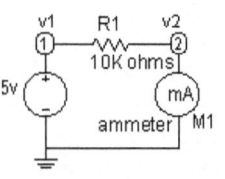
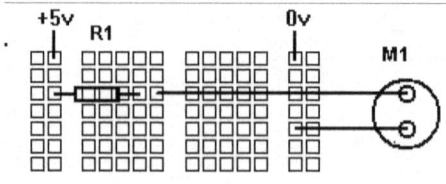

The Exponential Function The famous mathematician Leonhard Euler (1707-1783) defined the exponential function when he proved that the sum of a power series solution to a first order differential equation is his number e raised to the $-\alpha x$ power ($e^{-\alpha x}$) He also proved the property that the derivative of $e^{-\alpha x}$ is $-\alpha e^{-\alpha x}$. Consider a parallel RC circuit.

$$0 = \frac{1}{R}v(t) + C\frac{dv(t)}{dt} \qquad \Rightarrow \qquad -\frac{1}{RC}dt = \frac{dv(t)}{v(t)}$$

Integrate both sides using x as a dummy variable.

$$-\frac{1}{RC}\int_0^t dx = \int_0^t \frac{dv(x)}{v(x)} \qquad \Rightarrow \qquad -\frac{1}{RC}(t-0) = \ln v(t) - \ln v(0)$$

$$-\frac{1}{RC}t = \ln\frac{v(t)}{v(0)} \qquad \Rightarrow \qquad v(t) = v(0)e^{-\frac{t}{RC}}$$

Figure 712 *Plots of e^{-x} and $1-e^{-x}$*

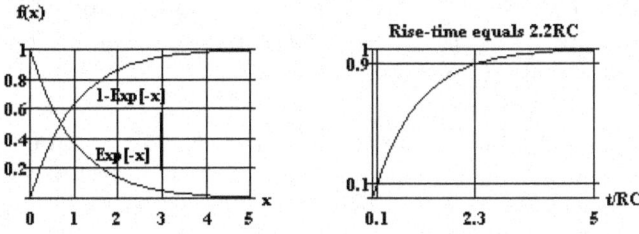

Step function u(t) The step function (Figure 713) is the implied forcing function when a switch is closed or opened.

Figure 713   *Definition of the unit step function*

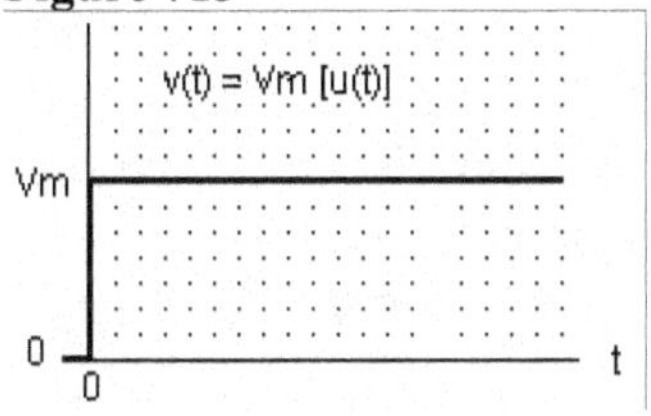

$$u(t) = \begin{cases} 0 & (t < 0) \\ undefined & (t = 0) \\ 1 & (t > 0) \end{cases}$$

Time Constant The step function transient response of circuits that reduce to one R and one L, or one R and one C takes the form of the exponential function exp(–x) and 1–exp(–x) where

$$x = \frac{t}{\tau} \quad and \ the \ time \ constant \ \ \tau = \frac{L}{R} \quad or \quad \tau = RC$$

Oscilloscope AC Voltage Measurements

AC Voltage AC voltage is jargon for the voltage magnitude of a sinusoidal signal voltage or any signal waveform on the screen.

Voltage Measurements Voltage is the electric potential, expressed in volts, between two nodes in a circuit. Usually one of these nodes is ground (zero volts). A basic way to measure a voltage is to count the number of divisions a waveform spans on the oscilloscope's vertical scale.

Circuit Connect the 50 ohm function generator output to the RC circuit input where R1 = 1K, C1 = (0.01+0.056) µF (Figure 714, page 33).

Sine Wave Connect the Channel A probe to the node whose AC voltage is to be measured. Select the function generator output waveform to be a sine wave with amplitude 2 volts peak to peak and frequency 1 kilohertz (1000 cycles per second). The period $T = 1/1000 = 1\times10^{-3}$ seconds (1 millisecond, 1ms)

Select AC as the input coupling. Adjust the channel *volts/div* control so that the AC signal is visible and spans many divisions. This will be the 0.5v/div setting since we know we are measuring 2v peak to peak. The AC signal now spans four ½volt divisions.

Adjust the channel *time/div* control to 1ms/div so that time across the 10 division screen is 10ms.

Select trigger source as *Channel* A. Select trigger type as *slope+*. Now comes the hard part that takes practice. Adjust the level control until the waveform is triggered at a waveform voltage of your choice. The result is a stable 10 cycle waveform display when frequency is 1,000Hz. This is preparation for a plot of amplitude versus frequency (Figure 71411 page 35)

Increase the frequency until the display amplitude shrinks to $4 \times 0.707 = 2.8$ divisions. This is −3dB frequency, which should be 10,000Hz.

7.2 Capacitor

A capacitor stores energy in the form of charge (electrons). The stored charge q in the capacitor C produces a voltage v across the two capacitor terminals where the determined by experiment q = Cv. Capacitors are formed by any two pieces of conducting material of any shape separated by a dielectric or free space. Consequently unwanted capacitors are everywhere and are connected to EVERY circuit node. These stray parasitic capacitors are a designer's demons. Their presence can be neglected if capacitor impedance z_C connected to a node is ten times greater than the impedance z_N at that circuit node. Since capacitor impedance equals $1/j2\pi fC$ they can be a serious problem at higher frequencies since the z_C impedance decreases as $1/f$ as frequency increases.

The unit of capacitance is the farad (F), which is a hugh capacitor. In the real world, capacitor values range from attafarads (10^{-15}F) in integrated circuits to the many microfarads (μF, 10^{-6}F) used in power supplies.

Theory The capacitor voltage-current vi relationship is found by differentiating $q = Cv$ to get $i = dq/dt = C\, dv/dt$. This is why current in a capacitor is proportional to the derivative of voltage rather than voltage. On the other hand, the voltage on a capacitor is proportional to the time integral of the input current so that $v = \int i\, dt\, /C = q/C$.

Figure 714

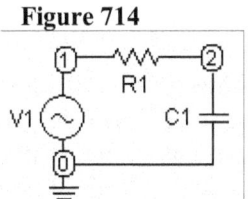

Theory: how to measure C The capacitor impedance $z_C = 1/pC = 1/j\omega C$ (Section 6.3) provides a means of indirectly measuring the capacitor's value by equating it to a resistor value. $R = |z_C|$ at what is referred to as the −3dB frequency. This is the frequency where an RC filter output v_2 (Figure 714) decreases by 3dB (decreases from 1 to 0.707).

$$\text{if } R = |z_C| \text{ then } R = \left|\frac{1}{pC}\right| = \left|\frac{1}{j\omega C}\right| = \frac{1}{\omega C} = \frac{1}{2\pi f_{-3dB}C} \Rightarrow C = \frac{1}{2\pi f_{-3dB}R}$$

The low pass filter R_1 and C_1 divide input voltage v_1.

$$v_1 = iR_1 + iz_C = iR_1 + i\frac{1}{j\omega C_1} = i\left(R_1 + \frac{1}{j\omega C_1}\right) \qquad v_2 = iz_C = i\frac{1}{j\omega C_1}$$

$$\frac{v_2}{v_1} = \frac{iz_C}{iR_1 + iz_C} = \frac{z_C}{R_1 + z_C} = \frac{1/j\omega C_1}{R_1 + 1/j\omega C_1} = \frac{1}{1 + j\omega C_1 R}$$

Theory: Capacitors in series The key observation for capacitors in series is that the same current flows through each capacitor in the series string. The voltage across each capacitor is Q_J/C_J. The total voltage is the sum of the individual capacitor voltages that is the sum of the Q_J/C_J terms. Since each Q is the integral of the same current i, all of the Q's are equal to each other. This means Q factors out of the voltage expression so that the total voltage equals Q times the sum of the $1/C_J$ terms or Q/C_{Total}. If the C's have different values the individual capacitor voltages are not equal even though the charges Q are equal.

A calculation of the sum of the voltage drops in the series string shows that the sum of the reciprocal values of capacitors in series can be replaced by one capacitor whose reciprocal value equals the sum. For capacitors in series "add reciprocals."

$$v(t) = \frac{1}{C_1} \int_0^t i\,dx + \frac{1}{C_2} \int_0^t i\,dx + \cdots + \frac{1}{C_n} \int_0^t i\,dx$$

$$v(t) = \left(\frac{1}{C_1} + \frac{1}{C_2} + \cdots + \frac{1}{C_n} \right) \int_0^t i\,dx = \frac{1}{C} \int_0^t i\,dx = \frac{Q}{C}$$

$$\frac{1}{C} = \frac{1}{C_1} + \frac{1}{C_2} + \cdots + \frac{1}{C_n}$$

Theory: Capacitors in parallel The key observation for capacitors in parallel is that the same voltage is across each capacitor. The sum of the currents flowing in the capacitors shows that the sum of values of capacitors in parallel can be replaced by one capacitor whose value equals the sum. For capacitors in parallel "add the values."

$$i(t) = C_1 \frac{dv}{dt} + C_2 \frac{dv}{dt} + \cdots + C_n \frac{dv}{dt}$$

$$i(t) = (C_1 + C_2 + \cdots + C_n) \frac{dv}{dt} = C \frac{dv}{dt}$$

$$\therefore C = C_1 + C_2 + \cdots + C_n$$

Theory: poles and zeros Circuit analysis of a circuit that can be analyzed by simplified lumped methods (6.3) produces expressions that are the ratio of two polynomials in the complex variable p. For example:

$$T(\omega) = \frac{v_2}{v_1} = \frac{\omega_0}{p + \omega_0} \qquad T(\omega) = \frac{v_2}{v_1} = \frac{p + \omega_2}{p^2 + p\omega_1 + \omega_0^2}$$ The zeros of a numerator are

zeros of $T(\omega)$. The zeros of a denominator are *poles* of $T(\omega)$.

Low Pass Filter Transfer Function $f_{-3dB} = 10,000Hz$ (Figure 71411)

$$T(\omega)_{low\_Pass} = \frac{v_2}{v_1} = \frac{z_c}{z_c + R_1} = \frac{1/j\omega C_1}{1/j\omega C_1 + R_1} = \frac{1}{1 + j\omega C_1 R_1} = \frac{1}{1 + j\omega/\omega_0} \text{ and } R_1 = \frac{1}{\omega_0 C_1}$$

$$C_1 = \frac{1}{\omega_0 R_1} = \frac{1}{2\pi f_{-3dB} R_1} = \frac{1}{2\pi \times 10^4 \times 10^3} = \frac{10^{-6}}{10 \cdot 2\pi} = 0.0159 \mu F$$

Practice: Construct the circuit in Figure 7141. Connect the function generator to the filter input at R_1. Let $R_1 = 1K$, $C_1 = (0.01 + 0.0056)\mu F$. Use an oscilloscope to measure f_{-3dB} (page 32).

Reminder: when evaluating results take into account the fact that the components have ±x% tolerance.

Figure 7141 RC Low Pass Filter

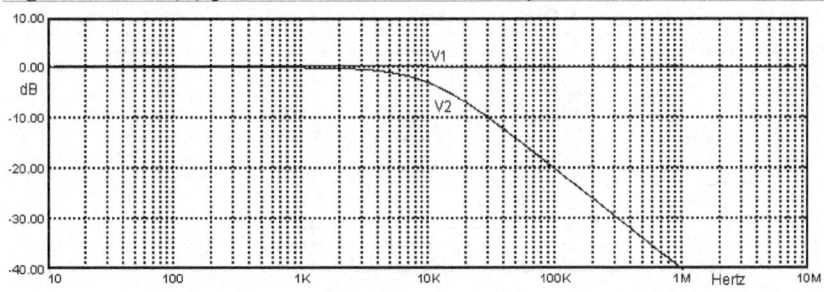

Spice Program 7141 Calculates Frequency Response

```
Fig7141.ckt factor 1/(p+a)
V1 1 0 AC 1 0  ; volts
R1 1 2 1000
C1 2 0 .0159155u
*.PLOT AC VDB(2) -40,10
.AC DEC 200 10 1e+007
.PLOT AC VP(2) -100,0
.PRINT AC VDB(2) VP(2) ;for numeric data
.TEMP 27
.end
```

Figure 714

Figure 71411 T(ω) plot where R=1K, C=0.0159 μF, –3dB f_0 = 10×10^3 Hz

Electronic Circuits – Practical Learning

The VDB4 plot in Figure 71421 below shows that the −3dB frequency doubles when capacitors are in series. This means the equivalent capacitor value is C/2. In other words the new capacitor reciprocal value is the sum of the reciprocal values of capacitors in series. The VDB3 plot shows that the −3dB frequency halves when capacitors are in parallel. This means the equivalent capacitor value is 2C. In other words the new capacitor value is the sum of the values of capacitors in parallel.

Figure 7142 Capacitors in series　**Figure 7143 Capacitors in parallel**

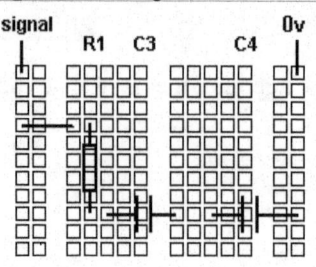 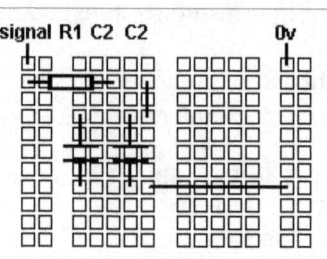

```
Fig7142.ckt        factor 1/(p+a)
V1 1 0   AC 1 0 sin(0 1 50K 0 0) ; volts
R1 1 2 1000
C1 2 0 0.0159155u
R2 1 3 1000
C21 3 0 0.0159155u
C22 3 0 0.0159155u
R3 1 4 1000
C3 4 5 0.0159155u
C4 5 0 0.0159155u
R5 5 0 100000000
.AC DEC 200 10 1e+007
.TEMP 27
.PLOT AC VDB(2) VDB(3) VDB(4) -5,0
.end
```

Figure 71421 T(ω) plots VDB2 C, VDB3 C+C (Parallel), VDB4 C/2 (Series)

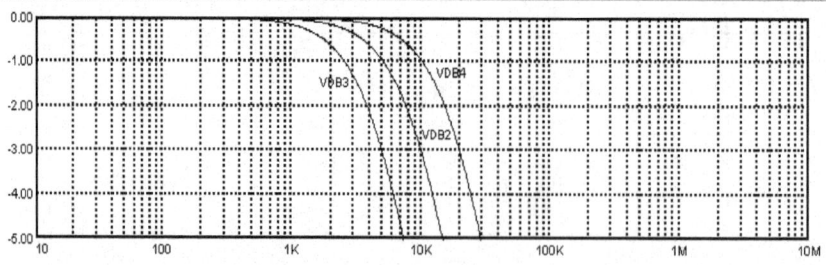

Theory: Steady State High Pass Filter If R and C are exchanged in Figure 714, then a steady state high pass filter results (Figure 715). C_1's impedance goes to infinity as frequency goes to zero, which results in attenuation of low frequencies and passing of high frequencies (Spice Figure 71511).

Figure 715

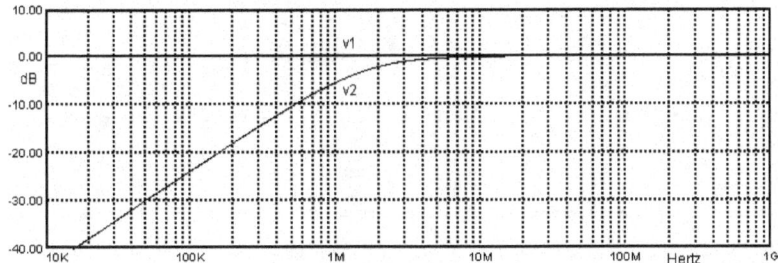

Practice Repeat the low pass filter practice.

Figure 71511 High Pass RC Circuit T(ω)=v_2/v_1

Transient State Transient behavior is produced in a circuit the instant anything changes such as when input signal(s) are connected to the circuit, when a waveform changes abruptly, or when circuit power is turned on. The transient waveforms that result depend on the kind of circuit components, their values and the input signal waveforms.

Figure 713 Step V_m u(t)

$v(t) = Vm [u(t)]$

Theory: A step function input signal (Figure 713) to the RC circuit (Figure 714) starts a transient event. Circuit solutions to transient events require solution of the circuit's time domain differential equation.

Heaviside provides the easiest solution (Section 6,3). We go directly to the transformed mesh equation where R and 1/pC are the R and C impedances.

For $\quad t > 0 \quad v_1(t) = v_m(t) = v_R(t) + v_C(t) = Ri(t) + \dfrac{1}{C}\int_0^t i(x)dx$

$$\rightarrow \quad v_1(p) = Ri(p) + \dfrac{1}{C} \times \dfrac{1}{p} i(p) = \left(R + \dfrac{1}{pC} \right) i(p)$$

37

Electronic Circuits – Practical Learning

$$v_1(p) = RI(p) + \frac{1}{pC}I(p) = \left(R + \frac{1}{pC}\right)I(p) \quad \text{let} \quad v(0^+) = 0$$

$$i(p) = \frac{1}{R + \dfrac{1}{pC}}v_1(p) = \frac{1}{1 + \dfrac{1}{pRC}}\frac{V_m}{pR} = \frac{1}{p + \dfrac{1}{RC}}\frac{V_m}{R}$$

Heaviside (and Laplace) showed that $1/(p + 1/RC) \rightarrow e^{-t/RC}$

$$i(t) = \frac{V_m}{R}e^{-at}u(t) \quad \text{where} \quad a = \frac{1}{RC}$$

In this RC circuit the exponential is $\exp(-t/RC)$. And, the frequency ω, time t relationship is

$$\boxed{\textit{Time constant } \tau = RC = 1/\omega_0.}$$

Practice Repeat the low pass filter practice. This time select a square wave waveform.

Spice program Calculates the RC Transient Response

```
Fig7152.ckt series RC circuit step
*PULSE( Vbase Vmax Tdelay Trise Tfall Twidth Tperiod )
V1 1 0 Pulse(0 0.9 0.2n 0n 0n 50n 100n)
R1 1 2 1000
C1 2 0 100f
.TRAN 1e-012 1e-009 0
.PLOT TRAN V(1) V(2) 0,1
.TEMP 27
.end
```

Figure 71521 Series RC circuit transient response to a step function where C= 100fF, R=1000 ohms → time constant RC=0.1×10⁻⁹ seconds

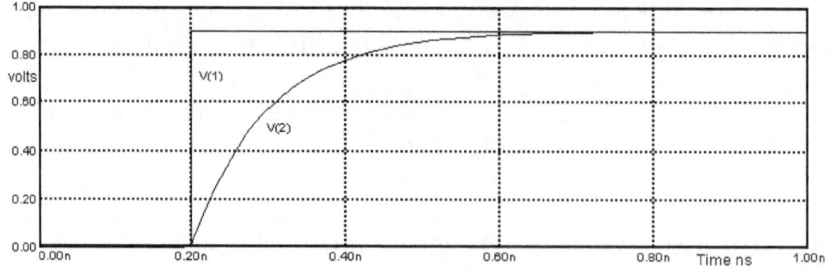

7.3 Inductor

An inductor stores energy in the form of current that is flowing. The typical inductor is wire wound into a coil. However know that *any* piece of *conducting* material in *any* form has the properties of an inductor. The magnetic field generated by the current flowing in the wire creates a counter-acting electric field which impedes changes to the current. This effect is known as Lenz's law and is stated mathematically as $v = L\, di/dt$. This is the voltage-current (vi) relationship of the inductor.

Consequently unwanted inductors exist in every circuit mesh. For example every wire is an inductor. The presence of these parasitic inductors can be neglected if their impedance z_L is say one tenth less than the impedance any circuit component z_M in series. Since inductor impedance equals $j\omega L$ parasitic inductors in series are serious problems at higher frequencies (Section 6.3).

The unit of inductance is the Henry (H) and practical inductor values range from nanohenries (nH, 10^{-9}H) to henries (H).

The voltage across an inductor is proportional to the derivative of current $v=L\, di/dt$. Alternatively, it can be said that the current in an inductor is proportional to the time integral of the voltage so that $i = \int v\,dt\,/L$.

Figure 716

Theory: How to measure an inductor The method is the same as the capacitor measurement method. The steady state inductor impedance (Section 6.3) $z_L = j\omega L$ provides a means of indirectly measuring the inductor's value by equating it to a resistor value. $R = |z_L|$ at what is referred to as the -3dB frequency. This is the frequency where an RL filter (Figure 716) output decreases by 3dB or 0.707.

$$\text{if } R =| z_L | \text{ then } R = \omega L = 2\pi f_{-3dB} L \;\Rightarrow\; L = \frac{R}{2\pi f_{-3dB}}$$

The low pass filter R_1 and L_1 divide input voltage v_1.
$$v_1 = iz_L + iR_1 = ij\omega L_1 + iR_1 = i(j\omega L_1 + R_1) \qquad v_2 = iR_1$$
$$\frac{v_2}{v_1} = \frac{iR_1}{iz_L + iR_1} = \frac{R_1}{z_L + R_1} = \frac{R_1}{R_1 + j\omega L_1} = \frac{1}{1 + j\omega L_1 / R}$$

Electronic Circuits – Practical Learning

Theory: Inductors in series

Inductors in series have the same current flowing through each inductor in the series string. Hence each inductor experiences the same di/dt. The induced voltage is proportional to L. Voltages in series add, and so do the inductance's. A calculation of the sum of the voltage drops in the series string shows that inductors in series can be replaced by one inductor whose value equals their sum.

$$v(t) = L_1 \frac{di}{dt} + L_2 \frac{di}{dt} + \cdots + L_n \frac{di}{dt}$$

$$v(t) = (L_1 + L_2 + \cdots + L_n) \frac{di}{dt} = L \frac{di}{dt}$$

$$L = L_1 + L_2 + \cdots + L_n$$

Theory: Inductors in parallel

Inductors in parallel have the same voltage across each inductor. A current entering a node to which inductors are connected in parallel divides among the inductors. The sum of the currents flowing in the inductors shows that the reciprocal of the sum of reciprocal values of inductors in parallel can be replaced by one inductor.

$$i(t) = \frac{1}{L_1} \int_0^t v(x)dx + \frac{1}{L_2} \int_0^t v(x)dx + \cdots + \frac{1}{L_n} \int_0^t v(x)dx$$

$$i(t) = \left(\frac{1}{L_1} + \frac{1}{L_2} + \cdots + \frac{1}{L_n} \right) \int_0^t v(x)dx = \frac{1}{L} \int_0^t v(x)dx$$

$$\frac{1}{L} = \frac{1}{L_1} + \frac{1}{L_2} + \cdots + \frac{1}{L_n}$$

Low Pass Filter Transfer Function f_{-3dB} = 1590Hz (Figure 71611)

$$T(\omega)_{low\_Pass} = \frac{v_2}{v_1} = \frac{R_1}{z_L + R_1} = \frac{R_1}{j\omega L_1 + R_1} = \frac{1}{1 + j\omega L_1 / R_1} = \frac{1}{1 + j\omega / \omega_0} \text{ where} R_1 = \omega_0 L_1$$

$$L_1 = \frac{R_1}{\omega_0} = \frac{R_1}{2\pi f_{-3dB}} = \frac{10^3}{2\pi \times 1590} = \frac{10^3}{10^4} = 10^{-1} = 100 \times 10^{-3} = 100mH$$

Practice: Construct the circuit in Figure 7161. Connect the function generator output signal to the free end of L_1. Let R_1=1K, L_1=100mH. Use an oscilloscope to measure f_{-3dB} (page 32). Adapt page 32 to measuring L.

Reminder: when evaluating results take into account the fact that the components have a $\pm x\%$ tolerance.

Figure 7161 RL Low Pass Filter

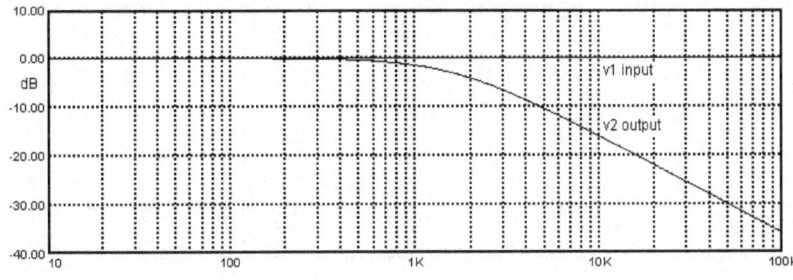

Spice Program 7161 - Low pass RL

```
Fig7161.ckt series RL circuit sine
V1 1 0 AC 1
L1 1 2 100m
R1 2 0 1000
*.PLOT AC VP(2) -100,0
.AC DEC 200 10 100000
.TEMP 27
.PLOT AC VDB(1) VDB(2) -40,10
.end
```

Figure 71611 Low Pass RL T(ω), R=1000ohms, L=100×10^{-3}H, f$_0$=1590Hertz

Electronic Circuits – Practical Learning

The VDB5 plot in Figure 71621 shows that the −3dB frequency halves when inductors are in series. This means the equivalent inductor value is 2L. In other words the new inductor value is the sum of the values of inductors in series. The VDB3 plot shows that the −3dB frequency doubles when inductor are in parallel. This means the equivalent inductor value is L/2. In other words the new inductor value is the reciprocal of the sum of the reciprocal values of inductors in parallel.

Figure 7162 Inductors **in parallel**   **Figure 7163** Inductors **in series**

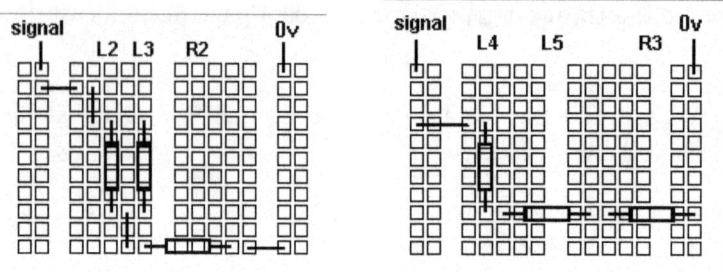

```
Fig7162.ckt   RL circuits    sine
V1 1 0 AC 1
L1 1 2 100m
R1 2 0 1000
L2 1 3 100m
L3 1 33 100m
Rx 33 3 1E-6        ; ignore
R2 3 0 1000
L4 1 4 100m
L5 4 5 100m
R3 5 0 1000
*.PLOT AC VP(2) -100,0
.AC DEC 200 10 100000
.TEMP 27
.PLOT AC VDB(2) VDB(3) VDB(5) -5,0
.end
```

Figure 71621 T(ω) plots VDB2 L, VDB3 L+L (Parallel), VDB5 L/2 (Series)

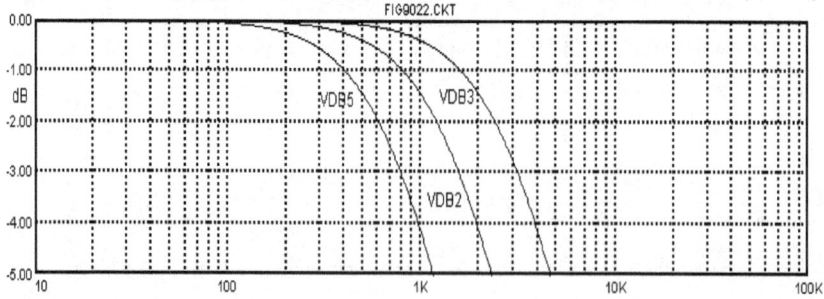

42

Theory: Steady State High Pass Filter If R and L are exchanged in Figure 716, then a steady state high pass filter results (Figure 717). L_1's impedance goes to zero as frequency goes to zero, which results in attenuation of low frequencies and passing of high frequencies (Figure 71711).

Figure 717

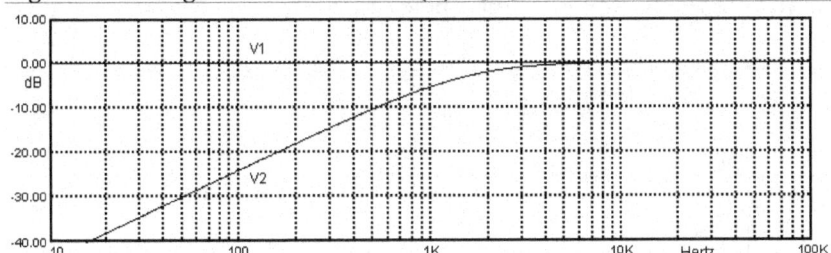

Practice Repeat the low pass filter practice.

Figure 71711 High Pass RL Circuit T(ω)=v₂/v₁

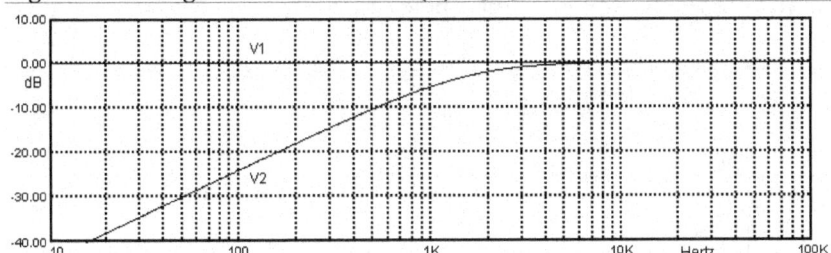

Transient State Transient behavior is produced in a circuit the instant anything changes – such as when input signal(s) are connected to the circuit, when a waveform changes abruptly, or when circuit power is turned on. The transient waveforms that result depend on the kind of circuit components, their values and the input signal waveforms.

Figure 713 Step Vₘ u(t)

$v(t) = Vm \, [u(t)]$

Theory: function input signal (Figure 713) to the RL circuit (Figure 716) starts a transient event. Circuit solutions to transient events require solution of the circuit's time domain differential equation.

At $t = 0$ *a step function steps* $v_1(t)$ *from 0 to* v_m *volts*

For $t > 0$ $v_1(t) = v_m = v_R + v_L = Ri(t) + L\dfrac{di(t)}{dt}$

Heaviside provides the easiest solution (Section 6,3). We go directly to the transformed mesh equation where R and pL are the R and L impedances.

Electronic Circuits – Practical Learning

$$V_1(p) = RI(p) + pLI(p) = (R + pL)I(p) \quad let \ \ i(0^+) = 0$$

$$I(p) = \frac{1}{R + pL}V_1(p) = \frac{1}{R + pL}\frac{v_m}{p} = \frac{v_m}{R}\left(\frac{1}{p} - \frac{L}{(R + pL)}\right)$$

$$i(t) = \frac{V_m}{R}[1 - e^{-\frac{R}{L}t}] \quad and \quad v_2 = Ri = v_m[1 - e^{-\frac{R}{L}t}] \quad where \quad \frac{1}{p} \to 1$$

In this RL circuit the exponential is exp(−Rt/L). And, the frequency ω, time t relationship is *Time constant* $\tau = L/R = 1/\omega_0$.

Practice Repeat the low pass filter practice.

Spice Program 7171 - High pass RL

```
Fig7171.ckt high pass RL circuit sine
V1 1 0 AC 1
R1 1 2 1000
L1 2 0 100m
*.PLOT AC VDB(1) VDB(2) -40,10
.AC DEC 200 10 100000
.TEMP 27
.PLOT AC VP(2) 0,100
.end
```

Figure 71711 High Pass RL Circuit T(ω)=v₂/v₁

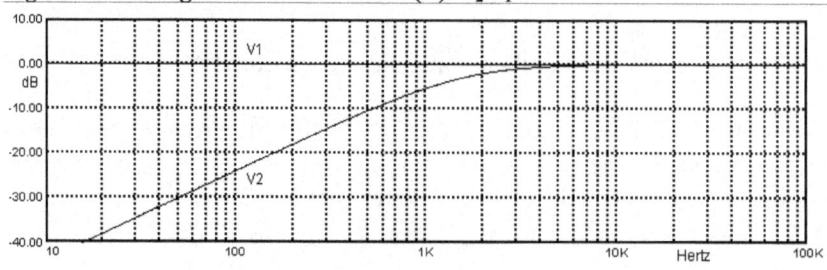

44

7.4 Transformer

The basic transformer is an assembly of two or more windings (coils of insulated wire) on a core. Each winding has self inductance L and a mutual inductance M with every other winding. Major properties of a transformer are conversion of voltage, and current to new levels with almost 100% efficiency when using high permeability magnetic cores. Consistent with conversion of voltage and current the associated impedances are *transformed* by the square of the turns ratio (n²) of the windings. Metallic connection between windings is not required. *There is no distinct mutual inductor you can touch.* Mutual inductance is a property of an assembly of two or more inductors that may simply be adjacent lengths of wire.

Figure 718 Flux in Windings

Theory: Coefficient of Coupling k The coefficient of coupling k relates the mutual M and L_1, L_2 inductances of two coils so that we do not have to be concerned about flux and flux linkages. The coupling between two coils can range from zero to 100%. Coupling is zero when there are no flux linkages from $coil_1$ to $coil_2$. Coupling is 100% when all of the flux generated by a current in one coil links the other coil. A proportionality constant (k) ranging from 0 to 1 is defined as the coefficient of coupling between two coils that is related to M, L_1, and L_2 (equation M). The coupled fluxes dictate a relationship between mutual and self inductance's based on the numbers of turns. Mutual inductance M_{12} is proportional to $N_1 \times N_2$, because current in the turns of one coil creates flux that cuts the turns of a second coil. Self inductance is proportional to N^2, because current in the N turns of a coil creates flux that cuts the same turns of the coil. Inductors L_1 and L_2 are proportional to N_1^2 and N_2^2.

$$v_1 = L_1 \frac{di}{dt} \quad \Rightarrow \quad L_1 = 10^{-8} N_1^2 \kappa = u_1 N_1^2$$

$$let \ L_1 = u_1 N_1^2 \quad and \quad L_2 = u_2 N_2^2 \quad and \quad M = M_{12} = M_{21} = u_M N_1 N_2$$

$$L_1 L_2 = u_1 u_2 N_1^2 N_2^2 = \frac{u_1 u_2}{u_M^2} M^2 = \frac{1}{k^2} M^2$$

$$M = k\sqrt{L_1 L_2} \ where 0 \le k \le 1$$

M is the actual mutual inductance, whereas $\sqrt{L_1 L_2}$ is the maximum possible value of mutual inductance.

Theory: Ideal Transformers Consider a two winding transformer whose windings have N_1 and N_2 turns, and the turns ratio n = N_2/N_1. The basic properties of ideal transformers are

1. Voltage is scaled by turns ratio (n)
2. Current is scaled by turns ratio (1/n)
3. Impedance is scaled by n^2
4. k=1 and M=$\sqrt{L_1L_2}$
5. No power is dissipated

Figure 719 Ideal transformer with load z_L

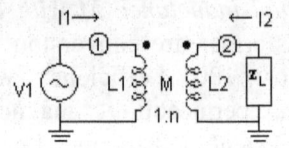

If $z_L=\infty$, then $I_2=0$ (Figure 719)

$$v_2 = pMi_1 \quad and \quad i_1 = \frac{v_1}{pL_1} \quad \Rightarrow \quad v_2 = pM\frac{v_1}{pL_1}$$

$$(7) \quad \frac{v_2}{v_1} = \frac{pM}{pL_1} = \frac{M}{L_1} = \frac{\sqrt{L_1L_2}}{L_1} = \frac{\sqrt{L_1L_2}}{L_1} = \sqrt{\frac{L_2}{L_1}} = \frac{N_2}{N_1} = n$$

If $z_L=0$, then KVL around the output mesh produces

$$(8) \quad 0 = pL_2i_2 - pMi_1 \quad \Rightarrow \quad \frac{i_2}{i_1} = \frac{pM}{pL_2} = \frac{N_1N_2}{N_2^2} = \frac{N_1}{N_2} = \frac{1}{n}$$

If $z_L \neq 0$ then $v_2=i_2z_L$.

$$(9) \quad z_{INPUT} = \frac{v_1}{i_1} = \frac{v_2}{n}\frac{1}{ni_2} = \frac{1}{n^2}\frac{v_2}{i_2} = \frac{z_L}{n^2}$$

> *The impedance z_L is transformed by the square of the turns ratio n. This is one reason why two or more windings coupled to each other are referred to as a transformer.*

Theory:Transformer Frequency Response T(ω) The ideal transformer transfer function T(ω)=1. However a practical transformer's T(ω) has a low frequency zero arising from shunt inductance kL, and a high frequency pole arising from *leakage* inductance (1–k)L in series arising from the fact coupling coefficient k does not equal one. In addition the coils as well as the core have finite resistance and parasitic capacitance.

Figure 720 A Transformer Equivalent Circuit

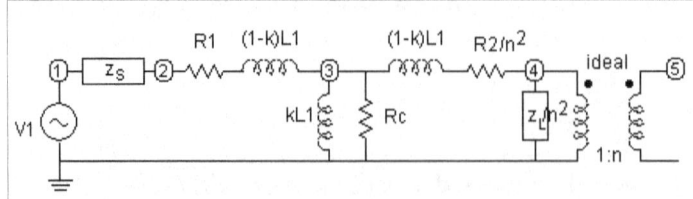

Theory: Low Frequency Circuit Analysis

At low frequencies the impedance of the series inductors is negligible compared to the source impedance. Capacitors are not included, because their impedance is large and negligible compared to the impedance of kL_1. This simplifies the equivalent circuit (Figure 721). Assume $R_L \ll R_C$ (Figure 720).

Figure 721 Low Frequency Model

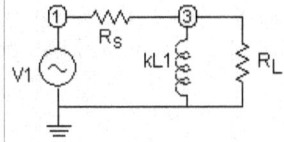

$$(14a) \quad 0 = g_s(v_3 - v_1) + \left(g_L + \frac{1}{pkL_1} \right) v_3 \qquad (14b) \quad g_s v_1 = \left(g_s + g_L + \frac{1}{pkL_1} \right) v_3$$

$$(14c) \quad T_L(p) = \frac{v_3}{v_1} = \frac{g_s}{\left(g_s + g_L + \frac{1}{pkL_1} \right)} = \frac{R_L}{R_s + R_L} \cdot \frac{p}{\left(p + \frac{R_s \parallel R_L}{kL_1} \right)}$$

$$T_L(p) = \frac{R_L}{R_s + R_L} \cdot \frac{p}{p + \omega_0}$$

If this transformer is in a 500 ohm line impedance system, then

$$(15a) \quad let \ R_s = R_L = 500\Omega, \ k = 0.99, \ L_1 = 1H$$

$$(15a) \quad f_0 = \frac{R_s \parallel R_L}{2\pi k L_1} = \frac{500 \parallel 500}{2\pi \cdot 0.99 \cdot 1} = 40.2 Hz$$

Figure 72111 Transformer Low Frequency Transmission (dB)

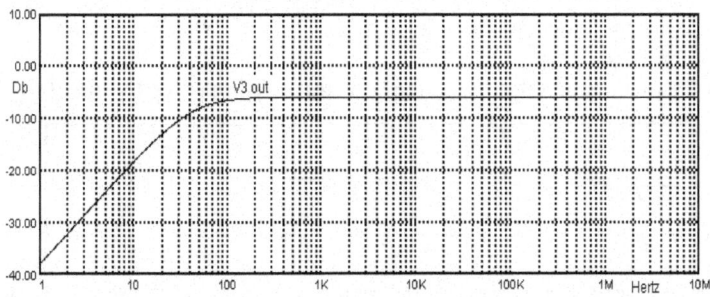

Figure 72112 Transformer Low Frequency Transmission (degrees)

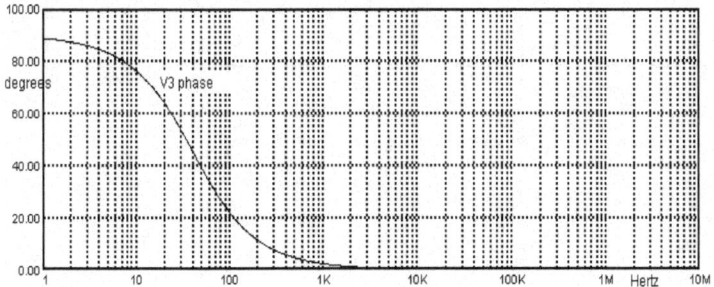

Electronic Circuits – Practical Learning

Determining the −3dB frequency in Figure 41011 is difficult, but Figure 72112 clearly shows the corresponding 45 degree phase at about 40Hz.

Theory: High Frequency Circuit Analysis The shunt kL_1 and R_C are omitted at high frequencies, because their impedances are very large compared to the load Z_L the source R and the leakage inductors' impedance in series. The windings

Figure722 High Frequency Model

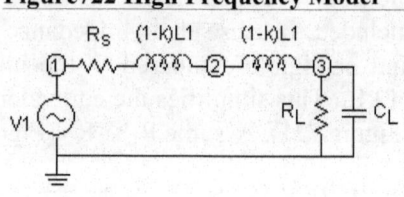

have capacitance that is usually represented by one capacitor C_L *as part of* Z_L (Figure 722).

$z_L=R_L$ Spice Program 7221 shows the effect of the windings' leakage inductance $(1-k)L_1$ with C_L omitted.

$(16a)$ $v_1 = [R_s + 2(1-k)pL_1 + R_L]i_1$

$(16b)$ $i_1 = v_1 \dfrac{1}{R_s + R_L + 2(1-k)pL_1}$

$(16c)$ $v_3 = i_1 R_L = v_1 \dfrac{R_L}{R_s + R_L} \dfrac{1}{1+\dfrac{2(1-k)L_1}{R_s + R_L}p}$

(17) $T_H(p) = \dfrac{v_3}{v_1} = \dfrac{R_L}{R_s + R_L} \cdot \dfrac{1}{1+\dfrac{p}{\omega_H}}$ *where* $\omega_H = \dfrac{R_s + R_L}{2(1-k)L_1}$

Figure 72211 Transformer High Frequency Transmission (dB), C_L omitted

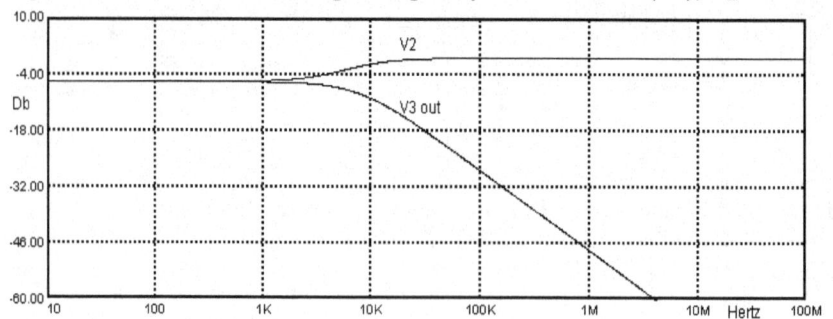

Theory: High and Low Frequency Circuit Analysis Traditional analysis uses low frequency and high frequency equivalent circuits (Figures 721 and 722).

Spice Program 7201

```
Fig7201.ckt   transformer xmsn T
V1 1 0 AC 1 0     ; volts
Rs 1 2 500
L1 2 0 1
L2 3 0 4
K12 L1 L2 0.99
RL 3 0 2000          ; 2000 here - 500 in prior plots
CL 3 0 500pf
*.PLOT AC VDB(2) VDB(3) -60,15
.AC DEC 200 1 1e+007
.TEMP 27
.PLOT AC VP(2) VP(3) -150,100
.end
```

Figure 72011 Transformer Transmission Function (magnitude dB)

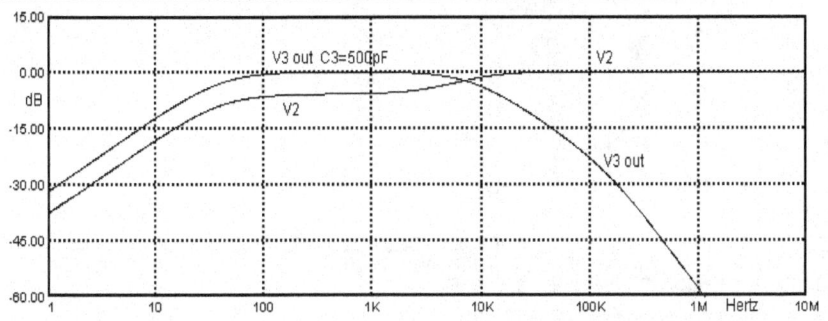

Electronic Circuits – Practical Learning

Practice: Obtain a transformer such as the Tamura TTC-108. Actually any transformer will do. Make measurements on your transformer as listed below.

Figure 723

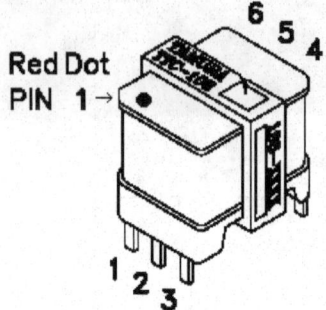

Red Dot
PIN 1→

Tamura TTC-108 Transformer Specifications

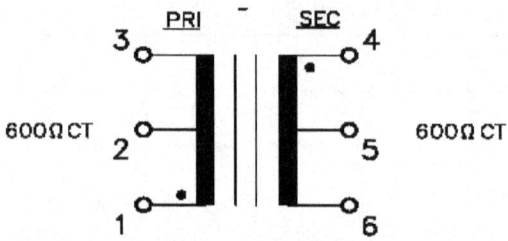

PRI   SEC

600Ω CT   600Ω CT

TELECOMMUNICATION DRY COUPLING TRANSFORMER DESIGNED TO OPERATE AT A MAX LEVEL OF +7dBm AND TO REFLECT A PRIMARY SOURCE IMPEDANCE OF APPROXIMATELY 600ΩCT WITH 600ΩCT LOAD ON SECONDARY

A. Electrical Specifications (@ 25 ° C)
1. Pri Source Impedance; 600Ω CT
2. Sec Load Impedance; 600Ω CT
3. Operating Level; −45 dBm to +7 dBm
4. Insertion Loss;
   1.4 dB MAX @ 1 KHz, 0 dBm
5. Frequency Response;
   ±0.5 dB 300 Hz to 3.5 KHz @ 0 dBm
6. Primary Impedance;
   600 Ω +15%, −5% @ 300 Hz to 3.5 KHz, 0dBm
   600 Ω +10%, −5% @ 500 Hz to 2.5 KHz, 0dBm
7. Longitudinal Balance;
   60 dB MIN @ 200 Hz to 1 KHz
   40 dBm MIN @ 4 KHz
8. DC Resistance;
   (1–3) = 44 Ω ±20%
   (4–6) = 56 Ω ±20%
9. Turns Ratio; (1–3) : (4–6) = 1 : 1.00 ±2%
10. Dielectric Strength;
   1500 Vrms 1 minute @ Pri to Sec, and Pri to Core
   1000 Vrms 1 minute @ Sec to Core
11. Total Harmonic Distortion;
   0.5% MAX @ 300 Hz to 3.5 KHz, 0 dBm

1 measure windings ohms Measure the transformer equivalent circuit components of a transformer. Measure resistance (*transformer* pins 1 to 3) R_1=56 Ω and (pins 4 to 6) R_2=77 Ω.

2 measure turns ratio n Ground *transformer* pins 3 and 6 (Figure 724, omit R_1). Adapt oscilloscope AC voltage measurements (page 32) to measure n. Connect signal generator V_1 to transformer pin 1. Connect AC coupled oscilloscope channel A and B inputs to transformer pins 1 and 4. Set frequency to 1KHz. Measure 1KHz AC voltages at pins 1 & 4.

Figure 721 Low Frequency Model	**Figure 724 Test Setup**
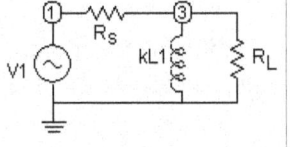	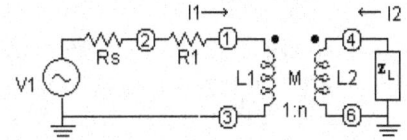

3 measure kL₁

Add R_1=470Ω to signal generator 50Ω R_S to make R_{total} about 650Ω including R_1, R_2 winding ohms (Figure 720). Connect 470Ω shown as R_1 in Figure 724. Connect 680Ω resistor to pins 4, 6. This is R_L. Omit C_L. Connect AC voltmeter input to pin 1.

Set frequency to about 1KHz. Adapt oscilloscope AC voltage measurements (page 32) to measure −3dB low frequency by reducing frequency gradually until scope shows 0.707=−3dB. Record the frequency. Use low frequency analysis to calculate kL_1. kL_1 is about 1H.

4 measure (1−k)L1

Signal generator R_S=50Ω. Add 470 Ω to make R_{total} about 650Ω as before. Set frequency to about 1KHz. Adapt oscilloscope AC voltage measurements (page 32) to measure −3dB frequency. Increase frequency gradually until screen shows voltage is 0.707=−3dB. Record the frequency. Use low frequency analysis to calculate kL_1.

Calculate k Given kL_1 and $(1-k)L_1$, calculate k.

Figure 722 Transformer High Frequency Model

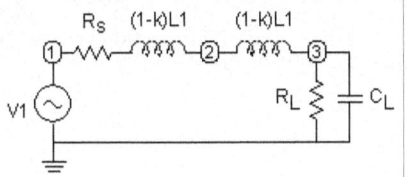

8 BJT Transistor Circuits

Transistors are active non-linear devices that amplify signals. One of the two major types of transistors is the bipolar-junction transistor (BJT). A current mirror, differential amplifier, LC tuned amplifier and LC oscillator are designed. Then inverter and NAND digital circuits are designed. The performance of all circuits is evaluated by Spice.

8.1 Bipolar Junction Transistor

Theory: A BJT has three terminals: emitter, base, and collector (Figure 801). There are two types of BJT's: npn devices that use electrons as the primary charge carrier and pnp devices that use "holes" as the primary charge carrier. (Figure 801 is not to scale). The regions n have a surplus of electrons, and regions p have a deficit of electrons referred to as a surplus of holes. We explain the *how*, but not the *why* of transistor functions. We leave the why of device physics to semiconductor texts.

Figure801 npn and pnp transistor symbols   **Figure 802 R Load**

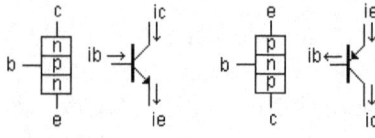
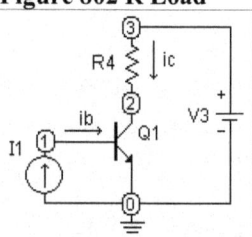

The bipolar junction transistor (BJT) is a current amplifying device. One form for current gain is expressed as $i_C = \beta i_B$ (Figures 801, 802). A graphical display of $i_C = \beta i_B$ is a two dimensional plot of y axis transistor collector current (I_C) and x axis collector-emitter voltage (V_{CE}) with I_B as a parameter. For example see the I_C/V_{CE} plot for $I_B = 54\mu A$ where β is about 160 (Figure 80311 page 54). Figure 80411 page 55 displays the typical BJT *vi constraint* with I_B as a parameter.

The concept of active device *loadline* superimposed on the vi constraint helps us to design bias circuits that establish I_C, the DC collector current, needed for a specific application (Figures 80312, 80411).

Theory Bipolar Junction Transistor vi constraint

An npn or pnp BJT may be described as *two diodes with a common p or n region* (Figure 801). The common p or n region implements current gain from base to collector, which is explained in semiconductor texts.

Amplification requires the *base-collector* diode to be reversed biased, and the *base-emitter* diode to be forward biased. The BJT $V_{BE}=V_1=0.7V$ is set by the current source I_1 forcing $i_B=I_1$ (Figure 803). The *base-collector* diode is reversed biased when the collector voltage $V_2>V_1$. The reversed biased *base-collector* diode is *not* off. It conducts current, because of transistor action. Similar comments apply to the pnp BJT.

Plotting a vi constraint A specific vi constraint is produced as follows. Let constant current I_1 fix the base current at 54µA. The vi constraint, collector current vs collector voltage v_{CE}, is plotted as the power supply voltage $V_2=V_{CE}$ is varied from 0 to 12 volts (Figure 80311 Spice program Fig8031).

Figure 803 npn transistor

With base current flowing the base-emitter voltage is about +0.7 volts. Therefore the base-collector diode is *forward* biased as the collector voltage V_2 increases from zero to 0.35 volts (Figure 80311). The BJT is *saturated* in this voltage range, where the current rises from 0 to about 8mA. The BJT is said to be *saturated* when the *base-collector diode is forward* biased. Increasing V_2 past $V_{ce}=0.35$ volts the BJT comes out of saturation as the base-collector diode becomes reversed biased to enter the *constant current active region*. In Figure 80311 I_C is not constant, because the transistor has a finite output resistance that draws current. The active region is used for linear amplification.

On-off operation Transistors in digital logical circuits are on or off. When on, the transistor is in the saturation region at point F_1 (Figure 80312 coordinate 1mA, 0.1V,). When off, the transistor is at point A (coordinate 0mA, 12V). This switching between points A and F_2 is represented by a *loadline*. If $R_4=4K$ then 3mA will flow when Q1 is on. This is a 4K loadline.

Figure 802 R Load

Spice Program 8031

```
Fig8031.ckt bipolar transistor vi constraint

.model 2N3904
+ NPN(Is=6.734f Xti=3 Eg=1.11 Vaf=74.03 Bf=416.4 Ne=1.259
+ Ise=6.734f Ikf=66.78m Xtb=1.5 Br=.7371 Nc=2 Isc=0 Ikr=0
+ Rc=1 Cjc=3.638p Mjc=.3085 Vjc=.75 Fc=.5 Cje=4.493p
+ Mje=.2593 Vje=.75 Tr=239.5n Tf=301.2p Itf=.4 Vtf=4
+ Xtf=2 Rb=10)

VCE 2 0 DC 0
I1 0 1 DC 54u
Q1 2 1 0 2n3904 ;c, b, e nodes 2, 1, 0

.DC LIN VCE 0 12 0.25    ;sweep VCE from 0 to 12 volts
.TEMP 27
.PLOT DC IC(Q1) 0,10M
.end
```

Figure 80311 BJT vi constraint

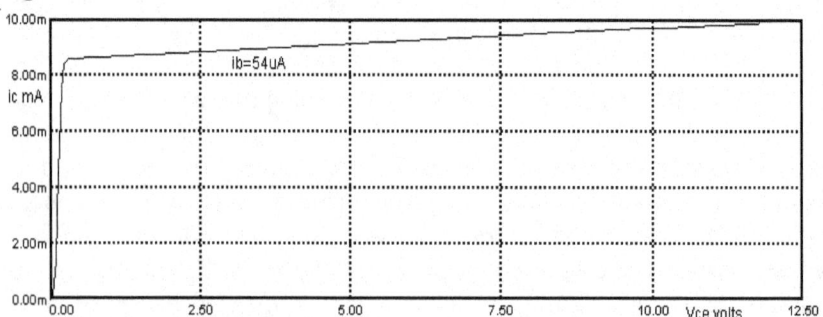

Figure 80312 BJT vi constraint with loadlines

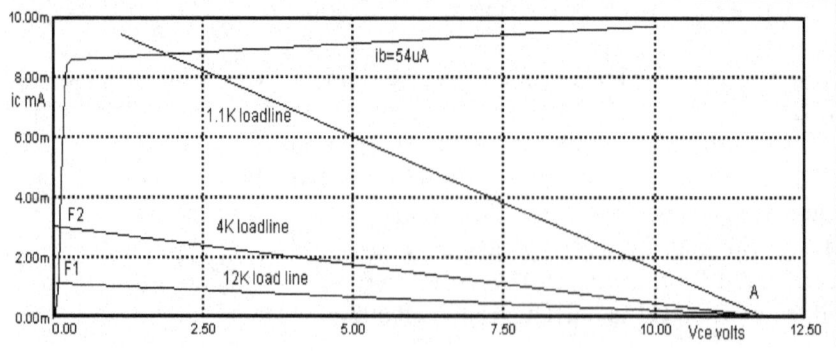

Spice Program 8041

```
Fig8041.ckt bipolar transistor

.model 2N3904
+ NPN(Is=6.734f Xti=3 Eg=1.11 Vaf=74.03 Bf=416.4 Ne=1.259
+ Ise=6.734f Ikf=66.78m Xtb=1.5 Br=.7371 Nc=2 Isc=0 Ikr=0
+ Rc=1 Cjc=3.638p Mjc=.3085 Vjc=.75 Fc=.5 Cje=4.493p
+ Mje=.2593 Vje=.75 Tr=239.5n Tf=301.2p Itf=.4 Vtf=4
+ Xtf=2 Rb=10)
```

```
I1 0 1 DC 0
* electron current flows into node 1
V2 2 0 DC 0      ; node 2 is positive
Q1 2 1 0 2N3904
.DC LIN V2 0 10 0.05 LIN I1 0 5e-005
1e-005
.TEMP 27
.PLOT DC IC(Q1) 0,10M
.PRINT DC IB(Q1)
.PRINT DC IC(Q1)
.end
```

Figure 804 npn vi constraint circuit

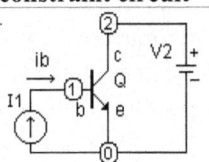

Figure 80411 2N3904 NPN Transistor vi constraint and loadline

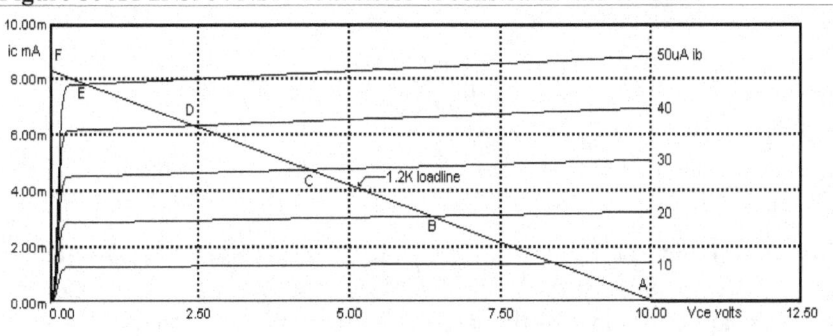

Spice A BJT's non linear vi constraint is analyzed by Spice programs. Spice program 8041 is in text file Fig8041.ckt that requires a BJT model written in the format shown. The 2N3904 BJT model was downloaded from a manufacturer's web site[1]. The model is in the form of a text file that was copied into Spice Program 8041 as *.model 2N3904.* (note the dot in .model). The program produces plots of the 2N3904 vi constraint (Figure 80411).

---

[1] http://www.fairchildsemi.com/

Collector current $I_C=0$ when base current $I_B=0$, and $V_{CE}=V_2=V_3=10V$. If $I_B=20\mu A$, then at node 2 $V_{CE}=6.3V$ when the load resistor R4 is $1.2K\Omega$ (Figure 802, and point B in Figure 80411). When I_B switches from $20\mu A$ to $40\mu A$, I_C switches from 3mA to 6.4mA, and the operating point travels along the loadline from point B to point D.

Figure 802 R Load

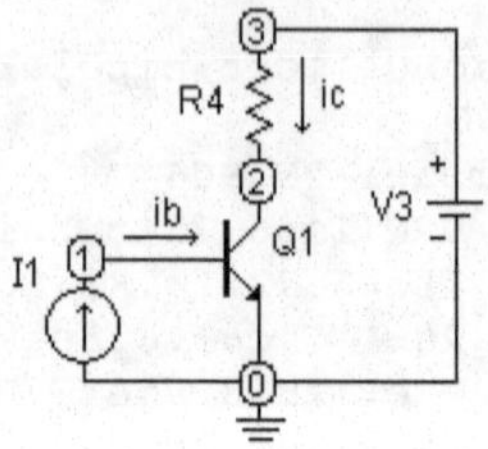

As the input I_B is incremented from 0 to $10\mu A$, $20\mu A$, $30\mu A$, etc, output I_C increments from 0 to about 1.4mA, 3.2mA, 5mA and so forth that indicates the current gain $\Delta I_C/\Delta I_B$ is about 1800/10=180 (device β). The operating point travels on the loadline from point A to points B, C, etc. as the input base current I_B increments. Point F is reached only if the transistor is a short, and then $I_C=8.33mA$. The get organized *saturation* region from 0 to about 0.35 collector volts is not entered by design in linear circuits.

Practice: BJT npn and pnp Transistors

Figure 801

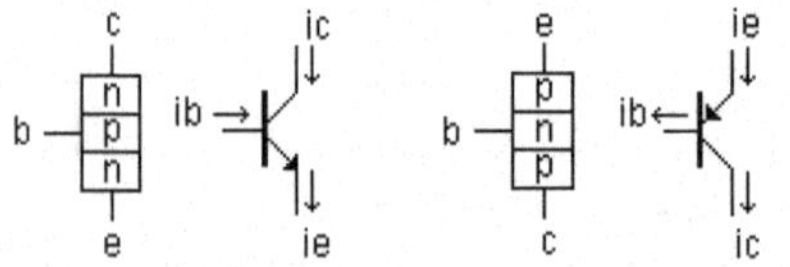

The ohmmeter The red lead of your ohmmeter is usually connected to the positive terminal of the internal battery. The black lead is connected to the negative terminal of the internal battery. We say usually, because your ohmmeter may be different. Please check.

Good or bad? Connect the positive ohmmeter lead to the collector, and the negative lead to emitter. The ohmmeter always reads infinity for a good transistor. This is so, because one diode is forward biased and the other is reversed biased no matter how the ohmmeter leads are connected. If the ohmmeter reads low ohms the transistor is defective. Repeat the process for a diode (+ to n, – to p)

npn or pnp? The transistor is an npn if the ohmmeter reads some number of ohms (the number depends upon the scale and the meter) when you connect the positive lead of your ohmmeter to the base (p of npn) and the negative lead to the collector (n of npn) or emitter (n of npn). If the ohmmeter reads infinite ohms reverse the leads, because the transistor is a pnp (or bad).

Figure P101	**Figure P102**

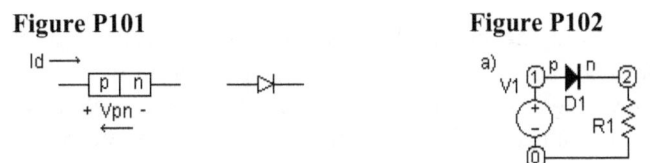

Transistor's 3 leads A metal case may have a tab (Figure P103).

Figure P103 Transistor lead assignments   **Figure P104 Pot**

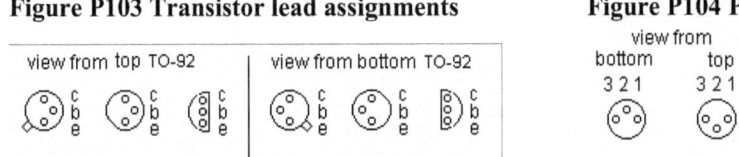

Transistor in a circuit At the next level a transistor is in a circuit and you wonder if it is OK. One way to find out is to short the base to emitter. This action turns off a good transistor. The collector voltage should now almost equal the supply voltage, except for any IR drop due to current supplied to the next stage. To make more collector current flow in the transistor add a resistor from V+=5v to the base (Figure P105). The collector voltage will decrease, & perhaps saturate the transistor, if the i_cR_1 drop is large enough.

Figure P105 Circuit Beta   **Figure P106 DC Bias**

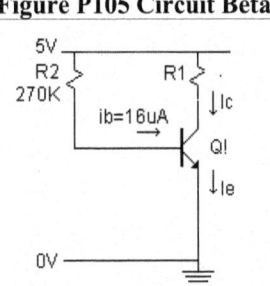
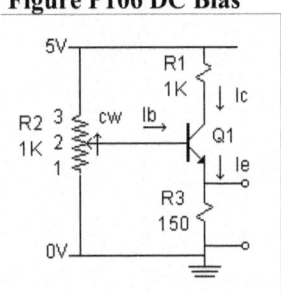

Electronic Circuits – Practical Learning

Circuit Beta Build the circuit on the breadboard (Figure P106). Use a 5V supply. Select Q_1 = 2N3904. Measure base current I_B. Let R_1 = 1KΩ, R2=1KΩ, R3=150Ω. Measure each R. Measure the voltage across each R. Calculate I_C and β_{ckt} .

FYI – Insert the 3 transistor leads in three *different* rows

DC Bias Circuit   Build the circuit P106 on the breadboard (Figure P107). Use a 5V supply. Select Q_1 = 2N3904. Adjust the pot arm so that *collector-to-ground* voltage is 3V. Measure base and emitter voltages. Calculate the collector current (2mA). What is the emitter current? (2mA+i_B)

Figure P107 P106 Circuit

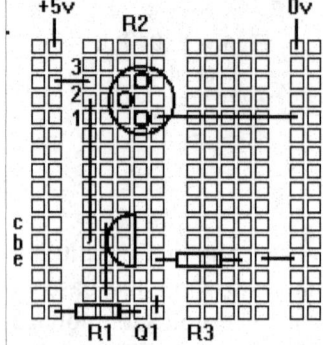

Connect to the base a 10KHz 0.5V peak to peak sine wave via 0.1μf capacitor. Measure the peak to peak sine wave voltage at the emitter and collector.

8.2 Small Signal T and Hybrid π Models

Model parameters are derived from BJT theoretical equations. BJT theory shows that in the active region the DC equations are

(1) $I_C = I_S e^{\frac{V_{BE}}{V_T}}$ $V_T = \dfrac{kT}{q} = 26mv$ $T = 300°K$

(2a) $I_E = \dfrac{I_C}{\alpha}$ (2b) $I_B = \dfrac{I_C}{\beta}$ (2c) $I_E = I_C + I_B$ (2d) $\beta = \dfrac{\alpha}{1-\alpha}$

Early voltage The Early voltage V_A is defined by projecting I_C plots to the left until they cross the zero current line (Figure 106). The 2N3904 model shows $V_A = V_{af} = 74.03V$. The Early voltage reveals finite output resistance.

(3) $I_{C\_Early} = I_S e^{\frac{V_{BE}}{V_T}} \left(1 + \dfrac{V_{CE}}{|V_A|}\right) = I_C \left(1 + \dfrac{V_{CE}}{|V_A|}\right)$

Figure 106 I_C finite slope defines Early voltage V_A

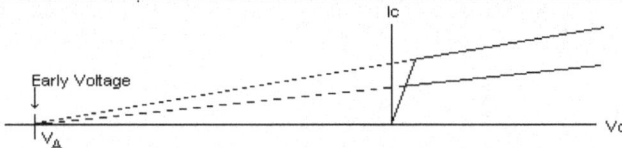

Small signal AC terms Each current and voltage is the sum of a DC and an AC term. Lower case subscripts indicate AC terms.

(4a) $v_{BE\_Total} = V_{BE} + v_{be}$ (dc + ac terms)

$I_C + i_c = I_S e^{\frac{v_{BE\_Total}}{V_T}} = I_S e^{\frac{V_{BE} + v_{be}}{V_T}} = I_S e^{\frac{V_{BE}}{V_T}} e^{\frac{v_{be}}{V_T}} = I_C e^{\frac{v_{be}}{V_T}}$

$e^x \approx 1 + x$ when $x = \dfrac{v_{be}}{V_T} < 1$

(4b) $I_C + i_c = I_C \left(1 + \dfrac{v_{be}}{V_T}\right)$ $\Rightarrow$ $i_c = \dfrac{v_{be}}{V_T} I_C$

Transconductance g_m (base to collector)

(5a) $i_c = \dfrac{I_C}{V_T} v_{be} \equiv g_m v_{be}$

(5b) $g_m = \dfrac{I_C}{V_T}$

8.3 *Classical* BJT amplifier

A classical BJT amplifier structure has a resistor in the emitter circuit (Figure 805) that makes this design relatively insensitive to changes in resistor values and transistor parameters. Resistor R_4 produces negative feedback that is a complex subject we take up later.

Figure 805

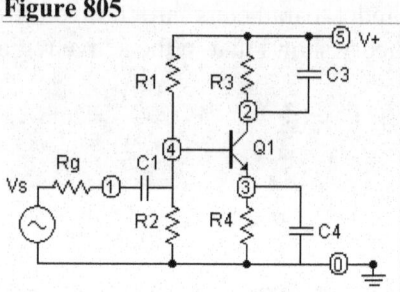

Bias Circuit – R1, R2, R3, R4

Selecting collector voltage $V_2 = 0.5V+$ places the DC operating point at the center of the BJT vi constraint (Figure 80411 page 55). $V_2 = 2.5v$ when $V+=5v$. Note that R_3 is the loadline R (Figure 80411).

(6) $V_2 = I_C R_3 = 2.5v$

R3 is the order of 10^3 ohms. This means R_3 is essentially the amplifier's output impedance, because the transistor output impedance r_0 is the order of 10^6 ohms

(7a) select $R_3 = 2700$ ohms

(7b) then $I_C = \dfrac{V_2}{R_3} = \dfrac{2.5}{2700} = 0.926\,mA$

The following decisions are not obvious. Select emitter voltage V_3=0.5V, because that reduces the V_2 swing range by only 0.5V. Given V_3 and I_C calculate R_4 (equation 8). The nearest standard 5% value for R_4 is 510 ohms, and base voltage V_4 is about 1.1V.

(8) $R_4 = \dfrac{V_3}{I_E} = \alpha\dfrac{V_3}{I_C} = \dfrac{\beta}{\beta+1} \times \dfrac{V_3}{I_C} = \dfrac{114}{115} \times \dfrac{0.5V}{0.926mA} = 535\Omega \approx 510\Omega$

(9) $V_4 = V_{be} + V_3 \approx 0.6 + 0.5 = 1.1V$

The R_1, R_2 voltage divider (Figure 805) defines V_4. Assume DC beta β=114. The base current $I_B = I_C/\beta = 926\mu A/114 = 8.12\mu A$. Arbitrarily select I_2=100μA >> 8μA. Observe that I_B flows in R_1.

(10) $R_2 = \dfrac{V_4}{I_2} = \dfrac{1.1}{100\mu A} = 11\,K\Omega$

(11) $R_1 = \dfrac{V_5 - V_4}{I_2 + I_B} = \dfrac{5 - 1.1}{(100 + 8.12)\,\mu A} = 36\,K\Omega$

Common Emitter Amplifier

The basic BJT amplifier is a common emitter amplifier (Figure 806).

Figure 806 T and Hybrid π Models - practical DC versions

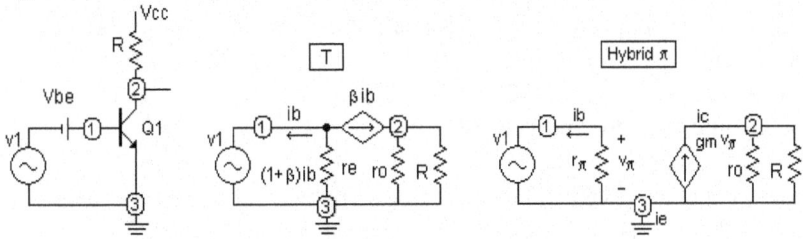

Small signal amplifier gain is calculated two ways to emphasize the T and Hybrid π models' properties. In the T model observe that i_E flowing in r_e is $i_B+\beta i_B$ (KCL). Let R approximate R in parallel with r_0.

Hybrid π Model

(12) $\quad v_2 = -i_c R = -g_m v_1 R \qquad$ and $\qquad v_1 = i_b r_{in} = i_b r_\pi$

(13) $\quad T_{v\pi}(0) = \dfrac{v_2}{v_1} = -\dfrac{g_m v_1 R}{v_1} = -g_m R = -\dfrac{I_c R}{V_T}$

T Model

(14) $\quad v_2 = -i_c R = -\beta i_b R \qquad$ and $\qquad v_1 = i_b r_{in} = i_b(1+\beta)r_e$

(15) $\quad T_{vT}(0) = \dfrac{v_2}{v_1} = -\dfrac{\beta i_b R}{i_b(1+\beta)r_e} = -\dfrac{\beta}{(1+\beta)}\dfrac{R}{r_e} = -\alpha\dfrac{R}{r_e} = -\dfrac{I_c R}{V_T} = -g_m R$

(16) $\quad T_{vT}(0) = -\dfrac{I_c R_3}{V_T} = -\dfrac{2.5}{26\cdot 10^{-3}} = 96$

Model parameters

Model parameters are derived from BJT theoretical equations. BJT theory shows that in the active region the DC equations are

(17) $\quad I_C = I_S e^{\frac{V_{BE}}{V_T}} \qquad V_T = \dfrac{kT}{q} = 26mv \qquad T = 300°K$

(18a) $\quad I_B = \dfrac{I_C}{\beta} \quad$ (18b) $\quad I_E = I_C + I_B = I_C + \dfrac{I_C}{\beta} = \dfrac{1+\beta}{\beta}I_C = \dfrac{I_C}{\alpha} \quad$ (18c) $\quad \beta = \dfrac{\alpha}{1-\alpha}$

Adding Capacitors

C_4 bypasses R_4 C_4 in effect shorts out the emitter resistor for maximum AC gain. Add C_4 in parallel with R_4 (Figure 805 page 60) to change R_4 to z_4. C_4 is a *bypass* capacitor, because when its impedance $1/\omega C_4$ is significantly less than R_4, very small AC currents flow in R_4. R_4 is in effect removed from the AC circuit. The pole and zero of impedance z_4+r_e become the zero and pole of $T_v(p)$.

A reactance chart reveals that if $C_4=30\mu F$, then its impedance equals 500 ohms at 10Hz. That is where we want the zero ω_z. As frequency increases past $f_z =10Hz$, C_4 starts to short R_4 out of the circuit, and gain begins to increase (Figure 80711). This is a process that completes as frequency increases past $f_P=193Hz$. At frequencies less than f_z $z_4 = r_e+R_4$, whereas at frequencies above f_P $z_4 = r_e$. The amplifier's 11dB gain at frequencies below f_z increases to 36.7dB at frequencies greater than f_P. The 25.7dB increase equals the ratio of $(r_e+R_4)/r_e$ in dB as well as the ratio f_P /f_z.

C_1 blocks DC C_1 disconnects source DC from input DC to pass only AC signals. The RC circuit corner is determined by selection of coupling capacitor $C_1=100\mu F$. The resistance from base to ground is $r_{in} = (1+\beta)(r_e+R_4)$ in parallel with R_1 and R_2. R_4 is in the circuit, because C_4 is out of the circuit at these low frequencies.

C_3 limits bandwidth Add C_3 to limit the high frequency response. Before we add C_3 we need to know the node 2 capacitance C_μ that is part of the transistor. We ran a Spice program with $C_3=0$ and learned the −3dB corner is about 14.4 MHz. So C_μ is about 3pf. Let $C_3 = 47+3 = 50pf$.

Figure 80711 Frequency Response $C_4=30\mu F$, $C_3=50pF$, $C_1=100\mu F$

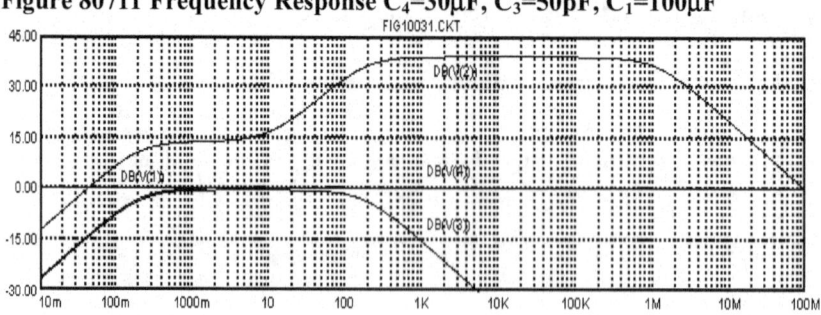

Write and run a Spice program representing the amplifier so that performance is evaluated (Figure 80711).

```
Fig8071 bjt ce amplifier with emitter resistor

.model 2N3904
+ NPN(Is=6.734f Xti=3 Eg=1.11 Vaf=74.03 Bf=416.4 Ne=1.259
+ Ise=6.734f Ikf=66.78m Xtb=1.5 Br=.7371 Nc=2 Isc=0 Ikr=0
+ Rc=1 Cjc=3.638p Mjc=.3085 Vjc=.75 Fc=.5 Cje=4.493p
+ Mje=.2593 Vje=.75 Tr=239.5n Tf=301.2p Itf=.4 Vtf=4
+ Xtf=2 Rb=10)


V5 5 0 DC 5
V1 11 0 AC 1  sin(0 0.2 10K 0 0)
Rs 11 0 10meg   ; spice rqd
Rg 11 1 0
C1 1 4 100u
R1 5 4 36K          ;base
R2 4 0 11K          ;base
Q1 2 4 3 2N3904     ;c b e
R3 5 2 2.7K         ;collector
C3 2 0 100p         ;collector
R4 3 0 510          ;emitter
C4 3 0 10u          ;emitter

.AC DEC 100 0.01 1E8
.PLOT AC VDB(1) VDB(4) VDB(2) VDB(3) -30,45
.PRINT AC VDB(1) VDB(4) VDB(2) VDB(3)
.TRAN 1e-008 0.0005 0 1e-008
.TEMP 27
.PLOT TRAN V(4) V(3) V(2) 0,5
.END
```

Spice program 8071 calculates the DC Voltages.

Node	Volts	Node	Volts	Node	Volts
5	5.000	11	0.000	1	0.000
4	1.114	2	2.622	3	452.502m

Electronic Circuits – Practical Learning

Practice: Build and Test the amplifier

Build the one stage amplifier (Figure 807) on the breadboard. Proceed as follows. Use a 5V supply. Then test it.

Figure 807 Classical Amplifier   **Figure 808 Classical BJT Amplifier**

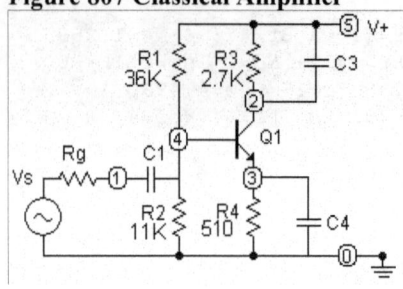
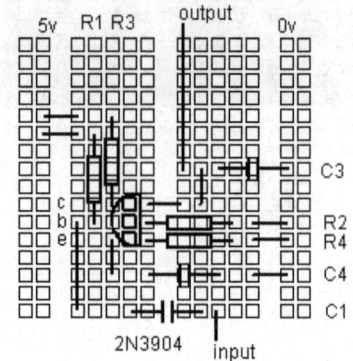

Insert Q1=2N3904.
Insert R_1, R_2, R_3, R_4 per Figure 807.

Measure the DC node voltages. Do they match voltages calculated by Spice as shown above?

Connect C_1=0.1µF to nodes 1, 4.
Omit C_3 and C_4 for now.

Is node 4 now at 1.11V? Measure base and collector currents. Calculate β_{ckt}.

Connect a signal generator (R_g is 50Ω) with V_S output to the node 1 input. *To isolate circuit node DC voltages from the AC voltmeter input connect a 1µF capacitor from node 2 (+) to the AC voltmeter input (−).* As you proceed use an oscilloscope to make sure sinewave signals are NOT distorted.

Measure the Classic Amplifier

1) Measuring frequency response of C_1=0.1µF.
Set signal generator sine wave frequency to about 10KHz. Adjust signal generator voltage so that the signal spans 5cm on the oscilloscope screen. Reduce frequency gradually until the signal spans 5cm × 0.89 (−1dB), and 5cm × 0.707 (−3dB). Record the 2 frequencies (375Hz, 200Hz). Is their ratio 2/1?

2) Measuring frequency response of $C_4=1\mu F$

Temporarily change C_1 to $100\mu F$. Set frequency to about 10KHz and a 5cm signal span. Reduce frequency gradually until signal spans 5cm $\times$ 0.707 (−3dB). Record the frequency, which should be about 193 Hertz.

Use equation 19b to calculate the theoretical −3dB frequency. This is f_z for pole. Work out a procedure for measuring f_z for zero. Measure the gain change with C_4 in or out. Compare to 25.7dB.

(19a) $\quad f_p = \dfrac{r_e + R_4}{2\pi R_4 r_e C_4} = \dfrac{(87+1.6K)10^6}{2\pi \cdot 1.6K \cdot 87 \cdot 10} = \dfrac{10^5}{2\pi \cdot 82.5} = 193Hz$

(19b) $\quad f_z = \dfrac{1}{2\pi R_4 C_4} = \dfrac{10^6}{2\pi(1.6K)10} = \dfrac{10^2}{2\pi(1.6)} = 10Hz$

3) Measuring frequency response of $C_3=1000\mu\mu F$.

Set the signal generator sine wave frequency to about 10KHz and a 5cm signal span. Increase frequency gradually until the signal spans 5cm $\times$ 0.89 (−1dB), and 5cm $\times$ 0.707 (−3dB). Record the 2 frequencies (55KHz, 100KHz). Is their ratio 2/1?

Use equation 20 to calculate the theoretical −3dB frequency for C_3. Compare to the measured value.

(20) $\quad f_0 = \dfrac{1}{2\pi(R_3 \| r_0)(C_\mu + C_3)} = \dfrac{1}{2\pi \cdot (6\|252.4)K\Omega \times 103.6pF}$

$\quad = \dfrac{1}{2\pi \cdot 5.86K\Omega \cdot 103.6pF} = 262KHz$

Electronic Circuits – Practical Learning

8.4 BJT Current Mirror

Constant current sources simplify bias circuits in discrete and integrated bipolar junction transistor circuits. The current mirror is a convenient way to produce constant current sources. Current mirror limitations are finite output impedance and operation over a range of voltages that is not rail to rail (V^+ to V^-). These limitations are not an issue in most circuits.

How it works The two transistor circuit qn_1, qn_2 as wired is the basic current mirror (Figure 809). I_{in} is set by R_1 and $I_{out}=kI_{in}$.

Figure 809 Current Mirror

$I_{C2}=I_{C1}$ because qn_2's V_{BE} equals qn_1's V_{BE}. However, base currents make $I_{out} < I_{in}$ (equations 22) since $I_{in}=I_{C1}+I_{B1}+I_{B2}$.

$$(22a) \quad I_{out} = I_{C2} = I_{C1} = I_{in} - 2I_B = I_{in} - 2\frac{I_{C2}}{\beta} = I_{in} - 2\frac{I_{out}}{\beta}$$

$$(22b) \quad I_{out} = \frac{\beta}{\beta+2}I_{in}$$

Design npn transistor qn_1 and resistor R_1 are connected in series from V^+ to V^- (Figure 809). The qn_1 transistor's base is connected to its collector so that $V_{BE1}=V_{CE1}$. The current in R_1 equals the qn_1 collector current plus the two base currents. For example, select $I_{in}=600\mu A$.

$$(23a) \quad R_1 = \frac{V^+ - V^- - V_{be}}{I_{in}} = \frac{1.8-(-1.8)-0.7}{600\cdot10^{-6}} = \frac{2.9}{600\cdot10^{-6}} = 4.833K$$

$$(23b) \quad \text{Let } R_1 = 4.7K \rightarrow I_{in} = \frac{2.9}{4.7\cdot10^3} = 617\mu A$$

Spice program 8091 calculates current mirror currents (Figure 80911 page 67). The slope of I_{C2} is produced by the finite output impedance. An estimate is as follows.

$$(24) \quad R_{out} = \frac{1v-(-1v)}{(635-618)\cdot10^{-6}} = \frac{2v}{17\cdot10^{-6}} = 117.6K$$

66

```
Fig8091.ckt  npn mirror

.model 2N3904
+ NPN(Is=6.734f Xti=3 Eg=1.11 Vaf=74.03 Bf=416.4 Ne=1.259
+ Ise=6.734f Ikf=66.78m Xtb=1.5 Br=.7371 Nc=2 Isc=0 Ikr=0
+ Rc=1 Cjc=3.638p Mjc=.3085 Vjc=.75 Fc=.5 Cje=4.493p
+ Mje=.2593 Vje=.75 Tr=239.5n Tf=301.2p Itf=.4 Vtf=4
+ Xtf=2 Rb=10)

Vdd   1 0 DC  1.8
Vss   4 0 DC -1.8

R1   1 2 4.7K
qn1  2 2 4 2N3904
qn2  3 2 4 2N3904
V3   3 0 DC 0


*.PLOT DC V(2) V(3) -1.8,1.8
.DC V3 -1.8 1.8 0.1
.TEMP 27
.PLOT DC I(R1) IC(QN1) IC(QN2) 600U,650U
.PRINT DC V(2) V(3)
.PRINT DC I(R1) IC(QN1) IC(QN2)
.end
```

Figure 80911 Current Mirror Currents R₁ = 4.7K

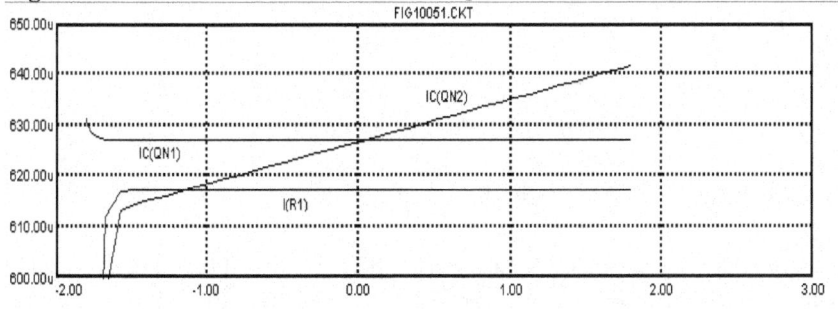

67

Electronic Circuits – Practical Learning

Practice: Build and Test the Current Mirror

Select I_{in}=0.8mA. Calculate R_1 (12K). Install R_1. Select V^+=5V, V^- = –5V, and 2N3904 for qn_1, qn_2. Build the current mirror circuit (Figure 811) on the breadboard (Figure 810). Place transistor and pot's pins on *3 different* rows.

Figure 809   **Figure 811**

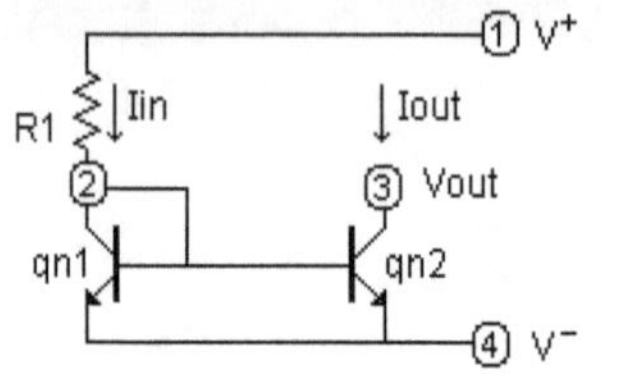

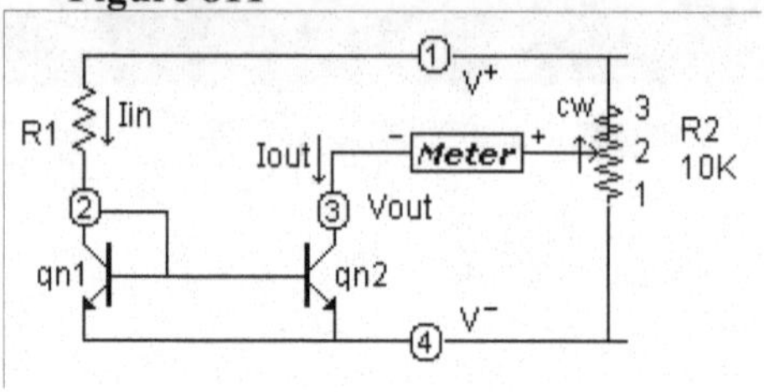

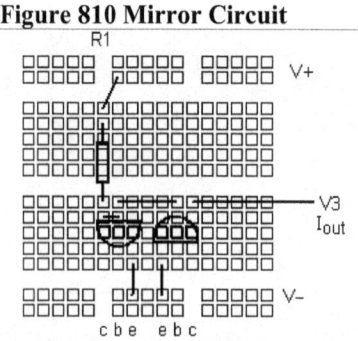

Figure 810 Mirror Circuit

Turn the pot CCW until the node 3 volts equal the node 2 volts.

Measure the DC volts at nodes 1, 2, 3, 4.

Make measurements that produce an I_{out} plot of $I_C(q_{N2})$ like Figure 80911 page 67.

Calculate r_{out} from the slope of the I_{out} plot (Figure 80911).

Set the node 3 DC volts equal to the node 2 DC volts. Measure the currents at nodes 2 and 3. Are they equal? (Equation 22b page 66)

8.5 Differential BJT Amplifier

A practical differential amplifier has high impedance inputs centered at 0 volts (ground), and a low impedance output. Positive V^+ and negative V^- power supplies allow for inputs centered at 0 volts. A BJT amplifier does not have high impedance inputs, nevertheless bandwidth and gain properties are advantageous. The emitter follower qn_3 at the output has relatively low output impedance. *The performance is set by transistor parameters.*

Transistors qn_1 and qn_2 are the basic differential amplifier with active load current mirror qp_1, qp_2. Emitter follower qn_3 has high input impedance that reduces the differential amplifier gain only by about 1dB. Current mirror R_1, qn_4, qn_5, qn_6, biases the circuit.

Figure 812 Differential Amplifier

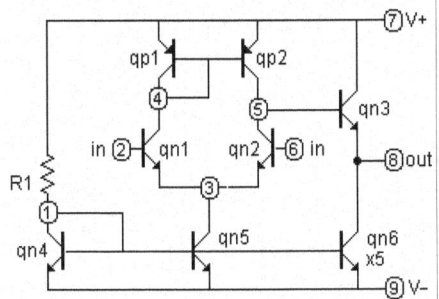

Select R_1 (Figure 812) to set the mirror current to I_1=0.6mA. Size qn_5=qn_4 so that I_5=0.6mA, and qn_6=$qn_4\times5$ so that I_6=3mA. Node 8 output impedance is equal to r_e plus $r_{outqn2}\|r_{outqp2}$ divided by $\beta+1$, because base current flows in r_{out} at node 5. The sum is the emitter follower source z. Transistor parameter values are taken from Spice program 8121 (page 71) numeric output and the transistor models.

$$(25)\quad R_o = r_{qn2} \| r_{qp2} = \frac{1}{g_{01}+g_{02}} = \frac{1}{\dfrac{1}{r_{01}}+\dfrac{1}{r_{02}}} = \frac{1}{\dfrac{I_{Cn}}{V_{An}}+\dfrac{I_{Cp}}{V_{Ap}}}$$

$$R_o = \frac{1}{\dfrac{0.3mA}{74}+\dfrac{0.3mA}{34}} = \frac{1}{\dfrac{1}{247K}+\dfrac{1}{113K}} = 77.7K\Omega$$

$$(26)\quad r_{in\_qn1} = 2r_\pi = 2\frac{\beta}{g_m} = 2\beta\frac{V_T}{I_C} = 2\cdot133\frac{26mV}{0.3mA} = 23.1K\Omega$$

$$(27a)\quad r_{out\_qn3} = r_e + \frac{r_{source}}{\beta+1} = \frac{\beta}{\beta+1}\cdot\frac{V_T}{I_C} + \frac{r_{qn2}\|r_{qp2}}{\beta+1}$$

$$(27b)\quad r_{out\_qn3} = \frac{133}{134}\cdot\frac{26mV}{1.65mA} + \frac{77.7K}{134} = 15.6+580 = 596\Omega$$

$$(28a) \quad r_{in\_qn3} = (1+\beta)(r_e + R_L) = (1+\beta)\left(\alpha\frac{V_T}{I_C} + R_L\right)$$

$$= 133\left(\frac{26mV}{1.65mA} + 4.7K\right) = 627K$$

$$(28b) \quad r_{load\_node5} = r_{in\_qn3} \parallel R_o = 627 \parallel 77.7 = 69.1K = 77.7 \times 0.89 \quad (-1.02dB)$$

The amplifier uses a 0.6mA current mirror, and an emitter follower, which loads node 5 with 627K for a 1.02dB loss. The amplifier drives 4.7K, and r_{out} of qn_6 is about 500 ohms. Compare 57.9dB gain with Spice 58.8dB.

$$(29) \quad R_1 = \frac{V^+ - V^- - V_{be}}{I_{ref}} = \frac{1.8 - (-1.8) - 0.7}{0.6mA} = \frac{2.9}{0.6 \cdot 10^{-3}} = 4.83K\Omega$$

$$(30) \quad I_1 = \frac{V^+ - V^- - V_{be}}{R_1} = \frac{2.9}{4.7 \cdot 10^3} = 0.617mA$$

$$(31a) \quad A_D = \frac{v_{out}}{v_{in}} = \frac{v_5}{v_1} = -g_m r_{out} = -\frac{I_C}{V_T} r_{out} = -\frac{0.297mA}{26mV} \cdot 69.1K\Omega = -789$$

$$(31b) \quad A_D = 20\log 789 = 57.9dB$$

Figure 81211 Differential Amplifier Frequency Response

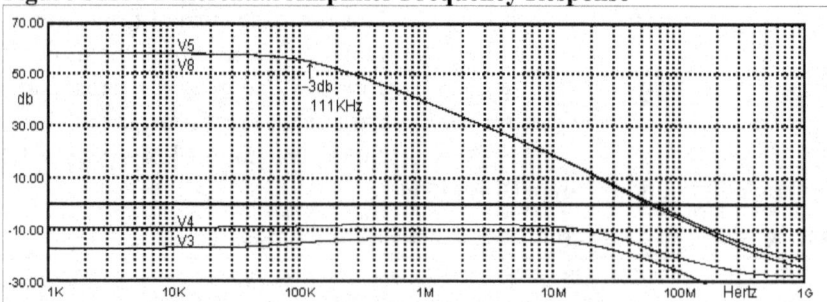

Figure 81212 Differential Amplifier DC Transfer function

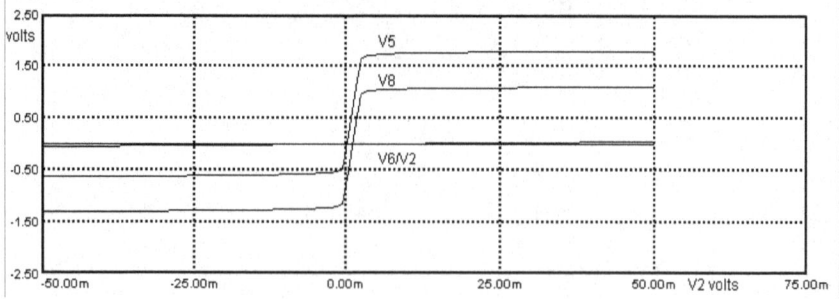

Spice program 8121

```
Fig8121.ckt  BJT diff amp
.model 2N3904
+ NPN(Is=6.734f Xti=3 Eg=1.11 Vaf=74.03 Bf=416.4 Ne=1.259
+ Ise=6.734f Ikf=66.78m Xtb=1.5 Br=.7371 Nc=2 Isc=0 Ikr=0
+ Rc=1 Cjc=3.638p Mjc=.3085 Vjc=.75 Fc=.5 Cje=4.493p
+ Mje=.2593 Vje=.75 Tr=239.5n Tf=301.2p Itf=.4 Vtf=4
+ Xtf=2 Rb=10)

.model 2n3906
+PNP(Is=455.9E-18 Xti=3 Eg=1.11 Vaf=33.6 Bf=204.7
Ise=7.558f
+ Ne=1.536 Ikf=.3287 Nk=.9957 Xtb=1.5 Var=100 Br=3.72
+Isc=529.3E-18 Nc=15.51 Ikr=11.1 Rc=.8508 Cjc=10.13p
+Mjc=.6993 Vjc=1.006 Fc=.5 Cje=10.39p Mje=.6931 Vje=.9937
+Tr=10n Tf=181.2p Itf=4.881m Xtf=.7939 Vtf=10 Rb=10)

V7 7 0 DC  1.8
V9 9 0 DC -1.8
V2 2 0 DC 0 AC 0.5
V6 0 6 DC 0 AC 0.5

qp1 4 4 7 2n3906
qp2 5 4 7 2n3906
qn1 4 2 3 2n3904
qn2 5 6 3 2n3904
qn3 7 5 8 2n3904
R1  7 1 4.7K
qn4 1 1 9 2n3904
qn5 3 1 9 2n3904
qn6 8 1 9 2n3904 3
R8 8 0 4.7K

.AC DEC 1000 1000 1e+009
.PLOT AC VDB(3) VDB(4) VDB(5) VDB(8) -30,70
.DC V2 -0.05 0.05 0.001
.TEMP 27
.PLOT DC V(2) V(6) V(5) V(8) -2.5,2.5
.PRINT DC V(8)
.end
```

Node	Voltage	Node	Voltage	Node	Voltage
7	1.800	9	-1.800	2	0.000
6	0.000	4	1.096	5	-139.204m
3	-634.237m	8	-819.921m	1	-1.147

MEGAHz	DB(V(8))	V(8)	
0.001	58.823	873.306	
0.111	55.820	618.037	-3dB

Electronic Circuits – Practical Learning

Practice: Build and Test the Differential Amplifier Build the differential amplifier circuit (Figures 812, 813) on the breadboard. Use ±5V supplies, 2N3904, 2N3906, R₁=12K.

Figure 812 Differential Amplifier

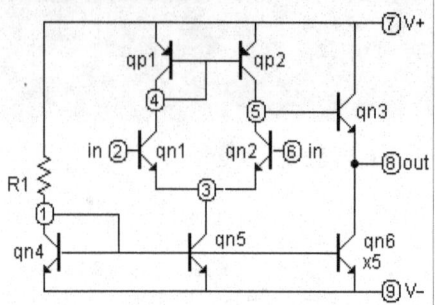

Ground input nodes 2 and 6. Measure DC volts at nodes 1, 3, 4, 5, 8 (Figure 812). Compare to DC operating point voltages (page 71) after increasing from 1.8 to 5 volts by a 5/1.8 scale factor.

Remove the ground from node 2. Connect the function generator output to node 2. Select sine wave output.

Figure 813

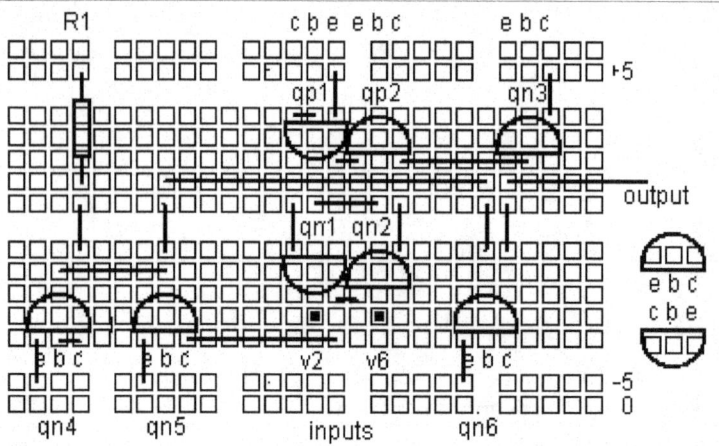

Make measurements that produce the equivalent of Figures 81211, 81212 page 70.

Connect the signal generator output and the oscilloscope channel 1 input to node 2. Connect the oscilloscope channel 2 input to node 8. Set the sine wave frequency to 1KHz. Adjust the signal input so that the signal spans 5cm (0dB) and output is not distorted

Measure the gain V_8/V_2 (1650) Compare to gain calculated by Spice (page 71) and Figure 81211 (page 70). Measure the frequency response −1dB and −3dB frequencies (25KHz, 50KHz). Compare to −3dB frequency calculated by Spice (Figure 81211 page 70).

8.6 LC Resonance

Electronic communications hardware is implemented by use of *frequency selective* circuits. Frequency selective circuits define *frequency bands* that confine the communications signals to those defined bands so that *communication channels* do not interfere with each other. Resonant circuits, *LC tuned circuits,* implement the frequency selective concept. (Many other types of circuits also implement the concept.) All frequency selective circuits are *filters* of various *bandwidths* and *frequency response* plots.

The most elementary frequency selective circuits are assemblies of one inductor L and one capacitor C that may be connected in series or in parallel (Figure 814). In both cases *resonance* occurs when the inductor and capacitor impedances are equal in magnitude at resonant frequency f_0.

$$(32a) \quad \omega_0 L = \frac{1}{\omega_0 C} \qquad (32b) \quad \omega_0^2 LC = 1 \qquad (32c) \quad \omega_0 = \frac{1}{\sqrt{LC}}$$

The impedance of a series LC circuit $z_s(\omega)$ (equation 33a) goes to zero when the frequency of the signal driving the circuit is the *resonant* frequency f_0, because at frequency f_0 the L and C impedances are equal in magnitude (equation 33b) and opposite in sign. At resonance node 2 is at zero volts (Figure 814). Note that a finite R_1 value does not change this result.

$$(33a) \quad z_s(p) = z_{series} = pL_1 + \frac{1}{pC_1} \qquad ideal \ L$$

$$(33b) \quad z_s(\omega) = \frac{1}{j\omega C_1} + j\omega L_1 = \frac{1}{j\omega C_1}(1 - \omega^2 L_1 C_1)$$

Figure 814 Series and Parallel Resonant circuits

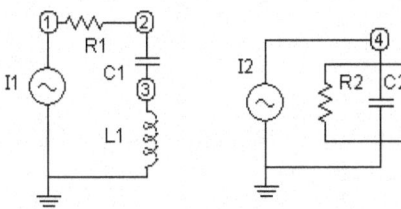

Electronic Circuits – Practical Learning

On the other hand the impedance of a parallel LC circuit $z_p(\omega)$, equation 34b, goes to infinity when the frequency of the signal driving the circuit is the *resonant* frequency f_0. At resonance node 4 could be at infinite volts (Figure 814). A finite R_2 value produces the finite voltage I_2R_2.

$$(34a)\quad z_p(p) = z_{parallel} = \cfrac{1}{\cfrac{1}{pL_2} + pC_2} \qquad R_2 = \infty$$

$$(34b)\quad z_p(\omega) = \cfrac{1}{\cfrac{1}{j\omega L_2} + j\omega C_2} = \frac{j\omega L_2}{(1-\omega^2 L_2 C_2)}$$

Finite resistor values alter the impedances as follows.

$$(35a)\quad z_s(p) = z_{series} = R_1 + pL_1 + \frac{1}{pC_1}$$

$$(35b)\quad z_s(\omega) = R_1 + j\omega L_1 + \frac{1}{j\omega C_1} = R_1 + \frac{1}{j\omega C_1}(1-\omega^2 L_1 C_1)$$

$$(36a)\quad z_p(p) = z_{parallel} = \cfrac{1}{\cfrac{1}{R_2} + \cfrac{1}{pL_2} + pC_2}$$

$$(36b)\quad z_p(\omega) = \cfrac{1}{\cfrac{1}{R_2} + \cfrac{1}{j\omega L_2} + j\omega C_2} = \cfrac{R_2}{1 - j\cfrac{R_2}{\omega L_2}(1-\omega^2 L_2 C_2)}$$

RLC circuit bandwidth is the number of Hertz between −3dB frequencies ω_2 and ω_1 of $z_p(\omega)$ that are above and below ω_0. At the −3dB frequencies the imaginary term magnitude equals 1 (equation 37b).

$$(37a)\quad \frac{R_2}{\omega_1 L_2}(1-\omega_1^2 L_2 C_2) = 1 \qquad (37b)\quad \omega_2 C_2 R_2 (\omega_2^2 L_2 C_2 - 1) = 1$$

$$(38a)\quad \frac{Q_p}{\lambda_1}(1-\lambda_1^2) = 1 \qquad (38b)\quad Q_p \lambda_2 (\lambda_2^2 - 1) = 1$$

$$Q_p = \frac{R_2}{\omega_0 L_2} = \omega_0 C_2 R_2 \qquad \lambda = \frac{\omega}{\omega_0}$$

Solution of the quadratics, and algebraic manipulations show the following for parallel or series resonant circuits.

$$(39a)\quad \frac{\omega_2 - \omega_1}{\omega_0} = \frac{1}{Q} \quad \Rightarrow \quad (39b)\quad BW = f_2 - f_1 = \frac{f_0}{Q}$$

Practice: Resonant Circuits

Inductor Model Create the model for a 1000µH inductor as follows.
Measure the ohms of the inductor (R=31 ohms).
Measure the self resonant frequency *srf* (f_{srf}=1.5MHz).
Given L and f_{srf} calculate the parasitic C (11.3pf)

Reactance Chart When L_1=1000µH and C_1=2700pF, a *reactance chart* shows that the LC resonant frequency (96.8 KHz) and characteristic impedance (608 ohms) values are vertical and horizontal lines through the point where the L_1=1000µH, and C_1=2700pF reactance lines cross.

Build the circuit Build the circuit (Figure 816) on the breadboard. Connect R_1=100 ohms to nodes 3 and ground, L_1=1000µH to nodes 2 and 3, and C_1=2700pF to nodes 4 and 2. (Node 1 is an internal source node.)

Figure 816 RLC Circuit   **Figure 815**

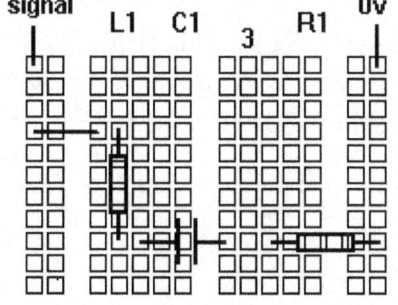

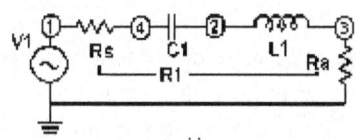

Steady State Response Connect V_1, the function generator (50 ohm R_S) to signal node 4. Select a 10KHz sine wave output.

Connect the oscilloscope to node 3. Adjust signal generator voltage so that the signal spans 5cm on the oscilloscope screen.

Increase the frequency slowly until you find a maximum at f_0, which is about 100KHz. Adjust the signal span so that it is 5 cm (0dB). At what frequency is v_2/v_1 a maximum? (104KHz) At what frequencies above and below f_0 does v_2/v_1=0.707 (−3dB, 88.9KHz, 120KHz)? What is the bandwidth (equation 39b, page 74)?

What is the total circuit resistance? Calculate $Q_S = \omega_0 L_1 / R_1$.

Electronic Circuits – Practical Learning

Transient Response of the circuit (Figure 815). R_a=100, L_1=1000µH, and C_1=2700pF

Figure 815

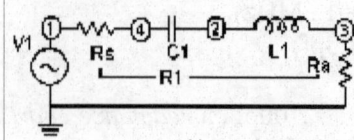

Find the characteristic impedance on the reactance chart (608 ohms). Calculate R=R_s+ R_a for Q=10, Q=1, Q=0.1(60, 600, 6000).

Set oscilloscope channel 1 to 1V/cm, channel 2 to 1V/cm.
Connect channel 1 to input node 4 (node 1 is not accessible), and channel 2 to output node 2.

Set 0V level at center screen. Select time base to be *10µs/cm so that full screen 10 cm sweep takes 100µs.*

Connect the function generator to node 4. On the Function Generator Set up a periodic pulse waveform with a 100µs period (10KHz), and 50µs pulse width. Adjust output pulse voltage to range from −1V to +1V.

For Q = 10, 1, and 0.1 (change R) observe that output V_3 has waveforms similar to Figure 81511.

Figure 81511

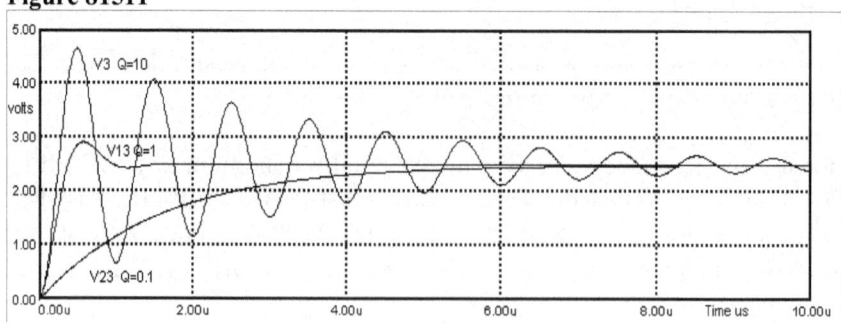

8.7 LC Tuned Amplifier

A tuned amplifier designed for use in an integrated circuit is a differential amplifier biased by a current mirror (Figure 817). Transistor capacitance affects the resonant frequency. It is found by plotting frequency response with 100pF C_1 out of the circuit leaving only 100μH L_1 in. Resonant frequency is 9.8MHz (Figure 81711), and C_P=2.64pF (equation

Figure 817 Tuned Amplifier

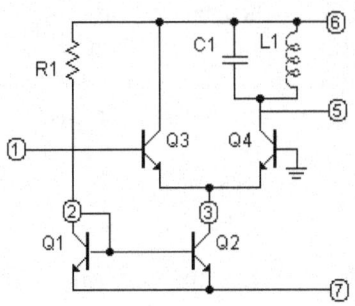

10). When 100pF is added the resonant frequency changes to 1.571MHz (Figure 81712).

$$(40) \quad C_P = \frac{1}{\omega_0^2 L} = \frac{1}{4\pi^2 (9.8 \cdot 10^6)^2 \cdot 100 \cdot 10^{-6}} = 2.64pF$$

$$(41) \quad f_0 = \frac{1}{2\pi\sqrt{L(C_P + C_1)}} = \frac{1}{2\pi\sqrt{100 \cdot 10^{-6} \times 102.64 \cdot 10^{-12}}} = 1.571MHz$$

Figure 81711 Tuned amplifier frequency response C_1=0

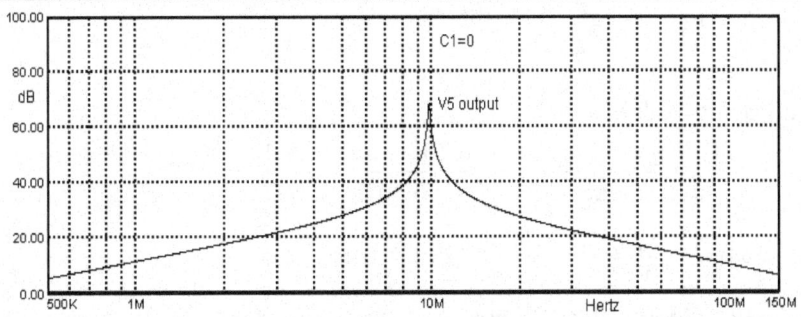

Figure 81712 Tuned amplifier frequency response C_1=100pF, R_2=250K

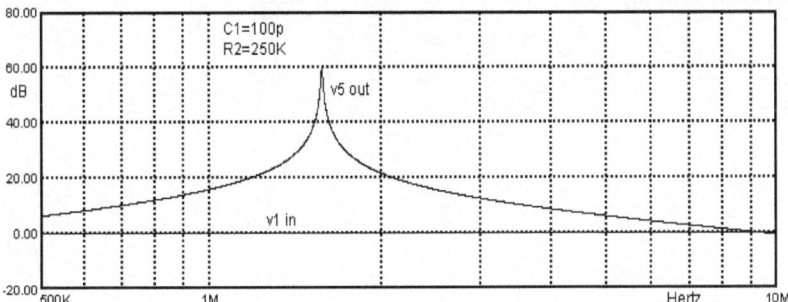

Electronic Circuits – Practical Learning

The AC schematic reduces to the Q_3 emitter follower driving common base amplifier Q_4 (Figure 818). Q_4's collector drives the *LC tank circuit*. BJT capacitors are omitted from the AC small signal equivalent circuit, because the poles they create are at significantly higher frequencies than the resonant frequency. The transmission function is derived from the AC equivalent circuit (Figure 819).

Figure 818 AC Schematic

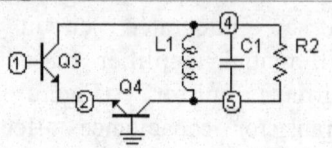

(42) $\dfrac{v_2}{v_1} = \dfrac{(1+\beta)r_e i_b}{(1+\beta)2r_e i_b} = \dfrac{1}{2}$

(43) $v_5 = \alpha i_e z_p(p) = \alpha z_p(p)\dfrac{v_2}{r_e} = \alpha z_p(p)\dfrac{1}{r_e}\dfrac{v_1}{2}$

(44) $T(p) = \dfrac{v_5}{v_1} = \dfrac{\alpha}{2r_e}z_p(p) = \dfrac{I_C}{2V_T}z_p(p)$

Gain at resonance (Figure 81712 shows 59.8dB).

(45) $T(\omega_0) = \dfrac{v_5}{v_1} = \dfrac{I_C}{2V_T}R_2 \parallel r_0 = \dfrac{0.3\text{mA}}{2 \cdot 26\text{mV}}250\text{K} \parallel 246\text{K} = 715.3 = 57.1\text{dB}$

MHz   v1   v5 dB (Spice 8171 data)
1.569 0.000   59.314
1.570 0.000   59.649
1.571 0.000   59.801   compare to 57.1 dB
1.571 0.000   59.749
1.572 0.000   59.501

Q_P and Bandwidth BW

(46) $Q_p = \dfrac{R_2 \parallel r_0}{\omega_0 L_1} = \dfrac{250 \parallel 246 \cdot 10^3}{2\pi \cdot 1.571 \cdot 10^6 \times 100 \cdot 10^{-6}} = \dfrac{124 \cdot 10^3}{987} = 125.6$

(47) $3\text{dB BW} = \dfrac{f_0}{Q_p} = \dfrac{1.571 \cdot 10^6}{125.6} = 12.51\,\text{KHz}$

Figure 819 AC Equivalent Circuit

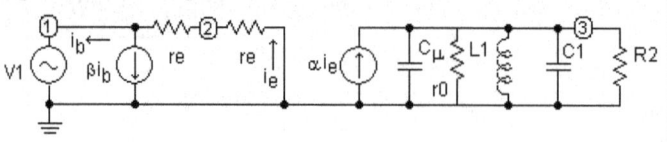

Spice Program 8171

```
Fig8171.ckt   npn tuned amplifier
.model 2N3904
+ NPN(Is=6.734f Xti=3 Eg=1.11 Vaf=74.03 Bf=416.4 Ne=1.259
+ Ise=6.734f Ikf=66.78m Xtb=1.5 Br=.7371 Nc=2 Isc=0 Ikr=0
+ Rc=1 Cjc=3.638p Mjc=.3085 Vjc=.75 Fc=.5 Cje=4.493p
+ Mje=.2593 Vje=.75 Tr=239.5n Tf=301.2p Itf=.4 Vtf=4
+ Xtf=2 Rb=10)

Vdd 6 0 DC 1.8
Vss 7 0 DC -1.8

R1 6 2 4.7K
qn1 2 2 7 2N3904
qn2 3 2 7 2N3904
V1 1 0 DC 0 AC 1 sin(0 0.01 1.5715E6 0 0) ; sine wave

qn3 6 1 3 2N3904
qn4 5 0 3 2N3904
R2 6 5 250K
L1 6 5 100u
*C1 6 5 100p

.TRAN 1e-008 2e-005 0 1e-009
.PLOT TRAN V(1) V(3) V(5) -1,5
*.AC DEC 5001 500000 1e+007
.AC DEC 5003 500000 1.5e+008.TEMP 27
.PLOT AC VDB(1) VDB(5) -20,80
.PRINT AC VDB(1) VDB(5)
.end
```

Practice: Build and Test the Tuned Amplifier   Build the circuit (Figure 817) on the breadboard.

Use ±5 Volt supplies. Select all Q=2N3904, L_1=1000µH and C_1=2700pF. The *reactance chart* shows that the LC resonant frequency (96.8KHz) and characteristic impedance (608 ohms) values are vertical and horizontal lines passing through the point where the L_1=1000µH, and C_1=2700pF reactance lines cross.

Let I_1=0.8mA. Calculate R_1 (12K). Install R_1 and measure I_1.

Calculate resonant frequency f_0. Calculate the characteristic impedance z_0. Compare your numbers to the reactance chart numbers.

Electronic Circuits – Practical Learning

Figure 817

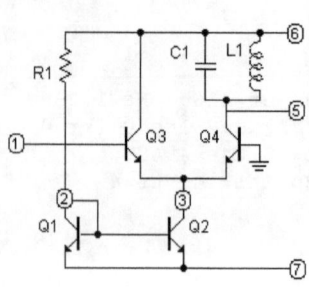

Figure 820

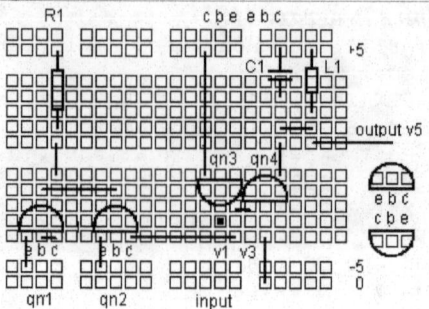

Steady State Response Connect the function generator's output to node 1. Select a 100KHz sine wave. Connect oscilloscope channel 1 to node 5. *Set oscilloscope channel 1 to 2V/cm.*

Set 0V level at center screen. Select time base to be *5μs/cm so that full screen 10 cm sweep takes 50μs.*

Adjust signal generator voltage so that the signal spans 5cm on the oscilloscope screen (0dB) and is an undistorted sine wave. You may have to reduce signal generator level as the frequency approaches resonance.

Adjust the frequency slowly until you find a maximum at resonant frequency f_0 (104KHz). Compare f_0 to 96.8KHz. At what frequencies above and below f_0 does $v_5/v_1=0.707$? (−3dB, 106, 101) What is the bandwidth? (5KHz) Q? (104/5) (equation 47 page 78)

Measure the gain at f_0. Calculate the gain (equation 45 page 78) and compare to your measurement. Work out the gain and Q differences.

Transient Response of the circuit you built.

On the Function Generator switch from sine wave to square wave and do not change frequency. What do you see at node 5? What did you expect to see?

8.8 LC Oscillator

An LC oscillator (Figure 821) is a tuned amplifier with feedback. The C_1LC_2 structure is a low pass filter. At the oscillation frequency there is a 180° phase shift through this structure, and transmission of a signal from transistor Q_4's base to collector produces another 180°

Figure 821 Colpitts Oscillator

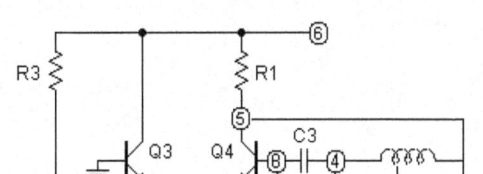

phase shift. The 360° shift satisfies one requirement for oscillation. The other requirement for oscillation is a return ratio T>1. Disconnect node 8 from Q_4's base, while assuming R_2 includes Q_4's input impedance. The return ratio T is the transmission from Q_4's base to node 5 to node 8. The upcoming circuit analysis is confirmed by Spice Figure 82121 page 83.

RC circuit R_2C_3 is a coupling circuit. At the oscillator frequency C_3 is a short so that $v_8=v_4$. Here are the node equations.

Figure 822 Colpitts Oscillator Model

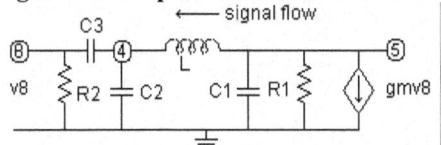

(48a) node 5 $\quad 0 = g_m v_4 - \dfrac{1}{pL}v_4 + \left(g_1 + pC_1 + \dfrac{1}{pL}\right)v_5 \quad$ note $v_4 = v_8$

(48b) node 4 $\quad 0 = \left(g_2 + pC_2 + \dfrac{1}{pL}\right)v_4 - \dfrac{1}{pL}v_5$

Solve 48b for v_5.

(49) $\quad v_5 = \left(1 + pLg_2 + p^2LC_2\right)v_4$

Substitute 49 into 48a.

(50) $\quad 0 = g_m v_4 - \dfrac{1}{pL}v_4 + \left(g_1 + pC_1 + \dfrac{1}{pL}\right)\left(1 + pLg_2 + p^2LC_2\right)v_4$

$\quad 0 = g_m + (g_1 + pC_1) + \left(g_1 + pC_1 + \dfrac{1}{pL}\right)\left(pLg_2 + p^2LC_2\right)$

81

(51) $\quad 0 = g_m + (g_1 + pC_1) + (g_1 + pC_1)(pLg_2 + p^2LC_2) + \dfrac{pLg_2 + p^2LC_2}{Lp}$

$\qquad 0 = g_m + (g_1 + pC_1)(1 + pLg_2 + p^2LC_2) + g_2 + pC_2$

$\qquad p = j\omega$

$\qquad 0 = g_m + g_1\left(1 - \omega^2LC_2\right) - \omega^2LC_1g_2 + g_2$

$\qquad\qquad\qquad + j\omega\left(g_1g_2L + C_1 + C_2 - \omega^2LC_1C_2\right)$

Imaginary part equals zero produces resonant frequency.

(52a) $\quad 0 = j\omega\left(g_1g_2L + C_1 + C_2 - \omega_0{}^2LC_1C_2\right)$

$\qquad \Rightarrow \ \omega_0{}^2 = \dfrac{C_1 + C_2 + g_2g_1L}{LC_1C_2}$

(52b) $\quad C_1 + C_2 + g_2g_1L = (100 + 100)\cdot 10^{-12} + \dfrac{100\cdot 10^{-6}}{3.3K\cdot 4.7K}$

$\qquad\qquad = (200 + 6.5)\cdot 10^{-12} = (206.5)\cdot 10^{-12}$

(52c) $\quad \omega_0{}^2 = \dfrac{C_1 + C_2 + g_1g_2L}{LC_1C_2} = \dfrac{206.5\cdot 10^{-12}}{100\mu H\cdot 100pF\cdot 100pF}$

$\qquad = 206.5\cdot 10^{12}$

$\qquad f_0 = \dfrac{\sqrt{\omega_0{}^2}}{2\pi} = \dfrac{\sqrt{206.5\cdot 10^{12}}}{2\pi} = 2.287MHz$

Figure 82121 shows about 8.7 cycles in 4µs, or 4/8.7µs period, and frequency = 8.7/4=2.175MHz.

Real part equals zero produces gain criteria. (Substitute 52a into 53a.)

(53a) $\quad 0 = g_m + g_1\left(1 - \omega_0{}^2LC_2\right) - \omega_0{}^2LC_1g_2 + g_2$

$\qquad g_mR_1 = \dfrac{C_2}{C_1} + g_2\left(\dfrac{L}{R_1C_1} + R_1\dfrac{C_1}{C_2} + \dfrac{L}{R_2C_2}\right)$

(53b) $\quad g_mR_1 = \dfrac{C_2}{C_1} + \left(\dfrac{L}{R_2R_1C_1} + \dfrac{R_1}{R_2}\dfrac{C_1}{C_2} + \dfrac{L}{R_2R_2C_2}\right)$

$\qquad = 1 + \left(\dfrac{100\cdot 10^{-6}}{3.3K\cdot 4.7K\cdot 100p} + \dfrac{4.7K}{3.3K} + \dfrac{100\cdot 10^{-6}}{3.3K\cdot 3.3K\cdot 100p}\right)$

$\qquad = 1 + (0.064 + 1.42 + 0.092) = 1.58 \quad \rightarrow \quad g_mR_1 > 1$

Spice Program 8212 Oscillator

```
Fig8212.ckt   tuned amplifier as oscillator

.model 2N3904
+ NPN(Is=6.734f Xti=3 Eg=1.11 Vaf=74.03 Bf=416.4 Ne=1.259
+ Ise=6.734f Ikf=66.78m Xtb=1.5 Br=.7371 Nc=2 Isc=0 Ikr=0
+ Rc=1 Cjc=3.638p Mjc=.3085 Vjc=.75 Fc=.5 Cje=4.493p
+ Mje=.2593 Vje=.75 Tr=239.5n Tf=301.2p Itf=.4 Vtf=4
+ Xtf=2 Rb=10)

Vdd   6 0 DC  1.8
Vss   7 0 DC -1.8

R1   6 2 4.7K
qn1  2 2 7 2N3904
qn2  3 2 7 2N3904

qn3  6 0 3 2N3904
qn4  5 8 3 2N3904

R2  6 5 3.3K
C1  5 0 100p
L1  5 4 100u
C2  4 0 100p
C3  4 8 1u
R3  8 0 5.6K ;100K

.TRAN 1e-010 1e-005 0
.TEMP 27
.PLOT TRAN V(3) V(8) V(5) -1.5,2.5
.end
```

Figure 82121 Oscillator Startup Waveforms

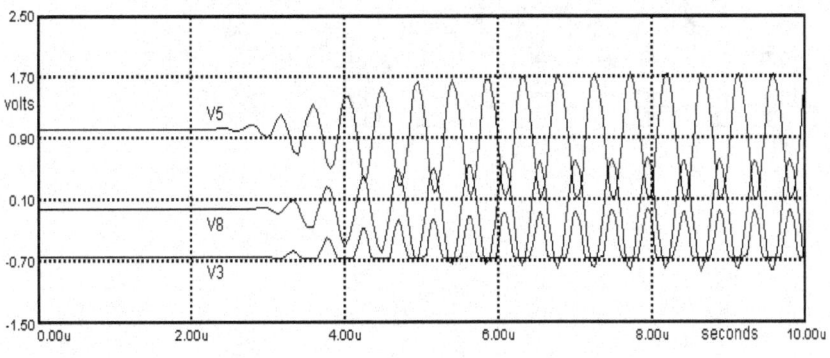

Electronic Circuits – Practical Learning

Practice: Build and Test the Colpitts Oscillator (Figures 821, 823).
Use ±5 Volt supplies. All are Q=2N3904.

R_3 12K $\quad$ R_2 5.6K $\quad\quad$ R_1 3.3K
C_1 680pF $\quad$ L_1 1000μH $\quad$ C_2 680pF $\quad$ C_3 1μF

Oscillates at 292KHz - 2V peak to peak.

Calculate the theoretical resonant frequency (equation 52 page 82).

Figure 821

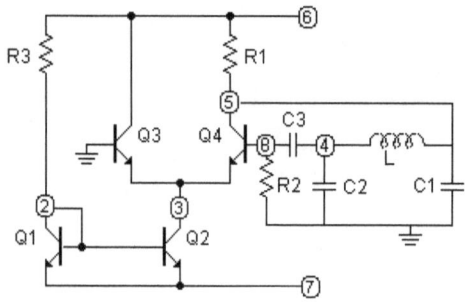

Figure 823

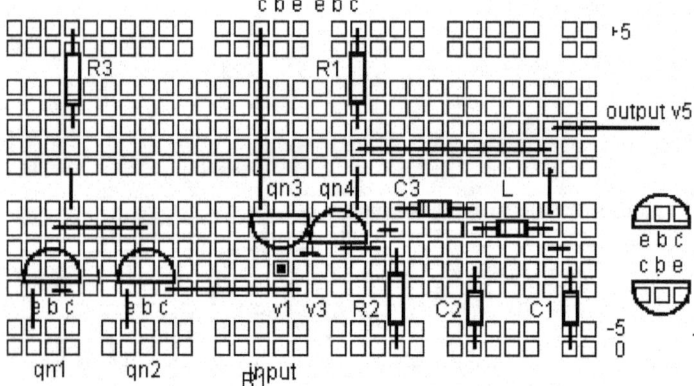

9 MOS Transistor circuits

The other major type of transistor is the MOS transistor. Circuits using MOS devices are a current mirror, differential amplifier, operational amplifier, LC tuned amplifier and LC oscillator, which are designed and their performance is evaluated by Spice. The circuits are based on two widely used analog circuits: the current mirror, and differential amplifier. They are the basic building blocks for analog integrated circuits. Then inverter and NAND digital circuits are designed. The performance of all circuits is evaluated by Spice.

9.1 MOS Transistor

The metal oxide semiconductor (MOS) transistor has two forms of the drain d, gate g, and source s regions (Figure 901 which is not to scale), which are nmos and pmos, We explain the *how*, but not the *why* of MOS transistors. We refer you to semi-conductor texts that explain the why of MOS device physics. (In modern MOS transistors the metal gate is replaced by a polysilicon gate.)

Figure 901 nmos & pmos transistors

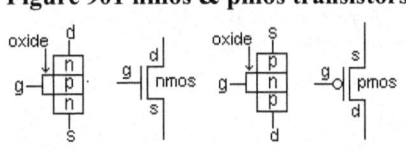

The principal difference from the BJT is that you *design* MOS devices to suit the application. The goal of the design is selection of values for channel length L, width W, and the correct value of mobility μ.

> MOS transistor design selects MOS *channel* length L and width W. Most MOS transistor circuits are intended to be *integrated circuits*. Circuit design includes layouts of the circuit suitable for manufacture as an *integrated circuit*.[1]

The MOS transistor is a *voltage to current* amplifying device. The g_m *transconductance* equation is $i_{DS}=g_m v_{GS}$. In effect, a graphical display of $g_m v_{GS}$ is a plot of transistor drain current (i_{DS}), drain-source voltage (v_{DS}) with gate-source voltage (v_{GS}) as a parameter (Figure 90321 page 87). This is another example of a *vi constraint*.

---

[1] N. L. Pappas, *CMOS Circuit Design - Analog, Digital, IC Layout*

Electronic Circuits – Practical Learning

The MOS transistor has three operating regions:
(1) the MOS is *turned off* when $v_{GS}<V_T$ the threshold voltage,
(2) the triode region is defined by $0<v_{DS}<V_{DSsat}=(v_{GS}-V_T)$,
(3) the (linear) saturation region is defined by $v_{DS}>V_{DSsat}$ (Figure 90321).

The MOS device is non-linear. Spice is designed to analyze MOS as well as BJT non-linear devices. The MOS Spice models used here are the University of California Predictive Technology Model Beta Version. The PTM model cards and BSIM models are available at

> http://ptm.asu.edu/
> http://www-device.eecs.berkeley.edu/bsim/

MOS Transistor vi constraint

An nmos or pmos transistor may be described as a *current channel connected to source s and drain d with a gate g over the channel that controls the electron or hole current flow in the channel* (Figure 902 which is not to scale). Zero DC current flows in the gate circuit of an

Figure 902 MOS structure

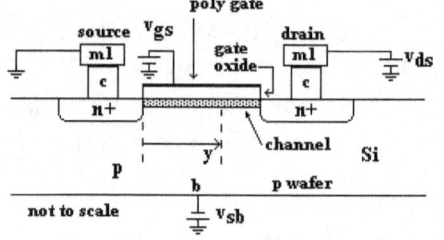

MOS transistor, because the gate to channel structure is a capacitor. This is a major difference from the DC BJT base current that flows. Channel length L is in the y direction. Channel width W is in the into the page z direction.

Plotting a vi constraint In an nmos circuit a positive V_{GS} allows electron current i_{DSn} to flow in the n-channel (Figure 903). A specific vi constraint is produced as follows. Maximum $V_{GS}=1.8V$ when a circuit is energized by a 1.8V battery. The vi constraint, drain current I_{DS} versus drain

Figure 903 nmos transistor

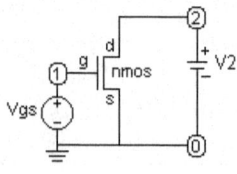

voltage V_{DS}, is produced for $V_{GS}=1.8V$ when power supply voltage $V_2=V_{DS}$ is increased from 0 to 1.8 volts (Figure 90311, Spice program Fig9031.ckt page 87).

Observe that there is no steep saturation region as there is in the BJT. The MOS is equivalent to a resistor in this region. This fact has many applications in chip layout.

Spice Program 9031

```
Fig9031.ckt    mos vi constraints
V2   2  0   DC 0
V1   1  0   DC 1.8
.include 180_N1P1.txt
MN1 2 1 0 0 N1 L=0.18u W=1.8u      ; 20/2=W/L
* (d g s sub nmos-model L  W)
.DC LIN V2 0 1.8 0.05
.TEMP 27
.PLOT DC ID(MN1) 0,1.5M
.PRINT DC ID(MN1)
.end
```

Figure 90311 N1 nmos Transistor Drain Current vi constraint

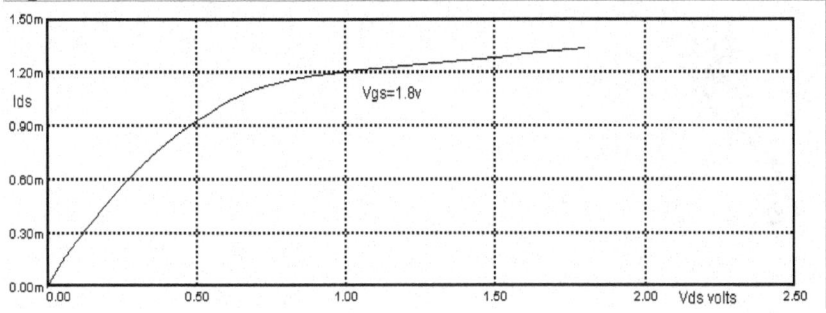

Spice Program 9032

```
Fig9032.ckt    mos vi constraints
V2   2  0 DC 0
V1   1  0 DC 0
.include 180_N1P1.txt
MN1 2 1 0 0 N1 L=0.18u W=1.8u      ; 20/2=W/L
.DC LIN V2 0 1.8 0.05 LIN V1 0.4 1.8 0.2
.TEMP 27
.PLOT DC ID(MN1) 0,1.5M
.end
```

Figure 90321 N1 nmos Transistor DC Drain vi Constraint

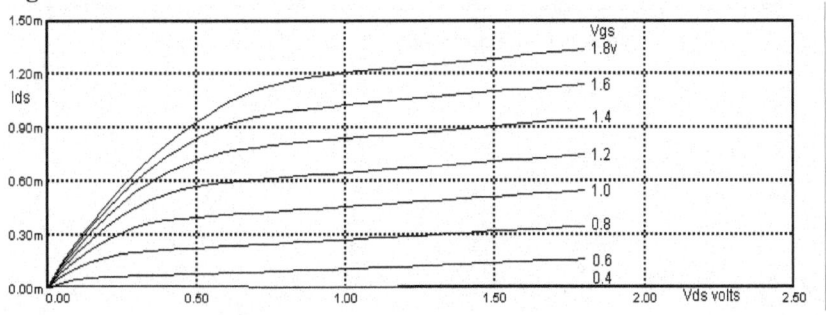

9.2 Available MOS Transistors

There are many discrete NMOS and PMOS high power transistors. On the other hand there are very few discrete low power transistors. The only transistors in a TO-92 three lead package that can plug into a solderless breadboard that we have found are the 2N7000 NMOS transistor, and BS250 and ZVP3306a PMOS transistors. (available at www.mouser.com).

Since detailed low current characteristics are not available guesses are necessary when, for example, calculating R_1 in a current mirror.

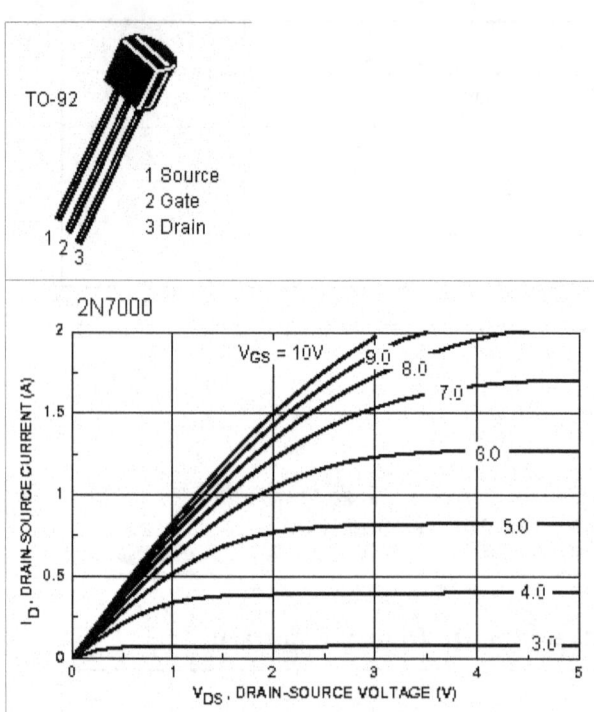

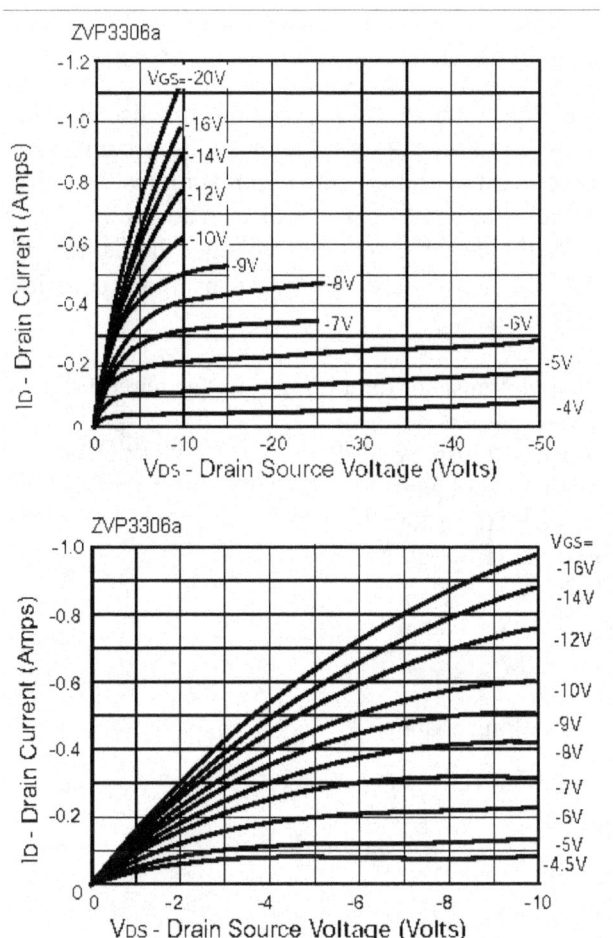

9.3 Current Mirror

Constant current sources have numerous applications in discrete and integrated MOS transistor circuits. The current mirror is a convenient way to produce constant current sources, while eliminating resistors that require relatively large chip areas. Current mirror limitations are finite output impedance and operation over a range of voltages that is not rail to rail (V^+ to V^-). These limitations are minimized by more complex mirror circuits not discussed here.

How it works at DC The two transistor circuit mn_1, mn_2 as wired forms a current mirror (Figure 904). I_{DS1} is set by R_1. Diode connected nmos transistor mn_1's drain is connected to the gate that forces V_{DS} to equal V_{GS}. Transistor mn_2's gate is connected to mn_1's gate so that $I_{DS_mn2} = $ constant $\times$ I_{DS_mn1} (equation 5). Equation 2 shows that if I_{DS} is selected, then V_{GS} is fixed.

Figure 904 Current Mirror

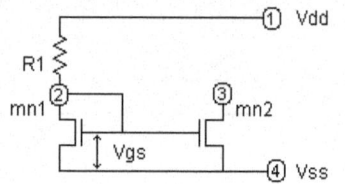

Saturation region equations, long channel

(1) $\quad I_{DSn} = \dfrac{1}{2}\mu_n C_{ox} \dfrac{W}{L}(V_{GS} - V_{Tn})^2 = \dfrac{1}{2}\beta(V_{GS} - V_{Tn})^2$

(2) $\quad V_{GS} = V_{Tn} + \sqrt{\dfrac{2 I_{DSn}}{\mu_n C_{ox}\dfrac{W}{L}}} = V_{Tn} + \sqrt{\dfrac{2 I_{DSn}}{\beta}}$

(3) $\quad R_1 = \dfrac{V_{DD} - V_{SS} - V_{DS}}{I_{DS}} = \dfrac{V_{DD} - V_{SS} - V_{GS}}{I_{DS}}$

The resistor is required, because the power supply voltage V_{DD}–V_{SS} is greater than V_{DS}. The output current I_{DS2} is proportional to I_{DS1}.

(4) $\quad I_{DS\_mn2} = \dfrac{1}{2}\mu_n C_{ox} \dfrac{W_2}{L_2}(V_{GS} - V_{Tn})^2 = \dfrac{W_2}{L_2}\dfrac{L_1}{W_1} \times \dfrac{1}{2}\mu_n C_{ox} \dfrac{W_1}{L_1}(V_{GS} - V_{Tn})^2$

$\quad I_{DS\_mn2} = \dfrac{W_2}{L_2}\dfrac{L_1}{W_1} \times I_{DS\_mn1}$

Design R_1 The value of the current mirror I_{DS} is specified by the circuit using the mirror. We want $L=5L_{min}$, because long L produces higher r_{out} Here is how W/L affects V_{GS}. Assume $I_{DS}=100\mu A$, W/L=10, $L=5L_{min}=10\lambda=0.9\mu m$.

$$(5a)\ V_{GS} = V_{Tn} + \sqrt{\frac{2I_{DSn}L}{(\mu_n C_{ox})W}} = 0.45 + \sqrt{\frac{2\cdot100\cdot10^{-6}\cdot10\lambda}{(200\cdot10^{-6})\cdot100\lambda}} = 0.45 + 0.316 = 0.77V$$

$$(5b)\ R_1 = \frac{V_{DD}-V_{SS}-V_{GS}}{I_{DSl}} = \frac{1.8-(-1.8)-0.77}{0.1mA} = \frac{2.83}{0.1\cdot10^{-3}} = 28.3K \approx 30K$$

Change to W/L=2 to increase V_{GS}

$$(6a)\ V_{GS} = V_{Tn} + \sqrt{\frac{2I_{DSn}L}{(\mu_n C_{ox})W}} = 0.45 + \sqrt{\frac{2\cdot100\cdot10^{-6}\cdot10\lambda}{(200\cdot10^{-6})\cdot20\lambda}} = 0.45 + 0.71 = 1.16V$$

$$(6b)\ R_1 = \frac{V_{DD}-V_{SS}-V_{GS}}{I_{DSl}} = \frac{1.8-(-1.8)-1.16}{0.1mA} = \frac{2.44}{0.1\cdot10^{-3}} = 24.4K \approx 24K$$

```
Fig9041.ckt   nmos mirror, L constant, W varied, R1=24K
.include 180_N1P1.txt
Vdd 1 0 DC 1.8
Vss 4 0 DC -1.8
R1 1 2 24K
mn1 2 2 4 4 N1 L=0.9u W=0.90u
mn2 3 2 4 4 N1 L=0.9u W=0.45u
mn3 3 2 4 4 N1 L=0.9u W=0.90u
mn4 3 2 4 4 N1 L=0.9u W=1.80u
V3 3 0 DC 0                      ; V3=Vds
.DC LIN V3 -1.8 1.8 0.1
.TEMP 27
.PLOT DC I(R1) ID(MN2) ID(MN3) ID(MN4) 0,250U
.PRINT DC I(R1) ID(MN2) ID(MN3) ID(MN4)
.end
```

Figure 90411 Mirror currents, for L=0.9um, W = 0.45, 0.9, 1.8μm, R_1=24K

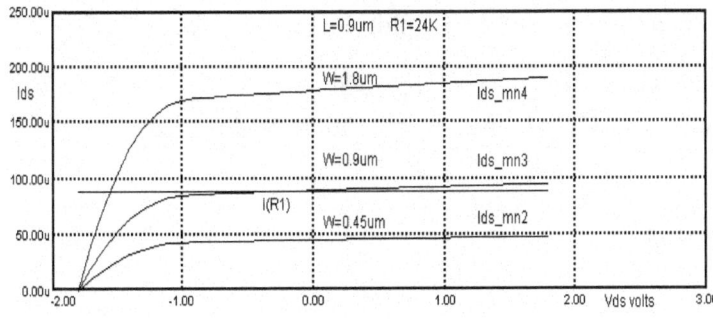

Electronic Circuits – Practical Learning

Practice: Build and Test the Current Mirror

The MOS transistors used here are the 2N7000 NMOS transistors. The models we have been able to find show L=100μm, W=100μm.

Select Vdd = 5V, Vss = −5V, and 2N2700 for qn₁, qn₂. Select I_{in}=0.8mA. Guess V_{CE} = 0.4V. Calculate R_1 (12K). Install R_1. Build the current mirror circuit (Figure 905) on the breadboard (Figure 906). Place a pot's pins on *3 different* rows.

Figure 904   **Figure 905**

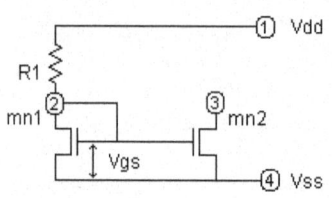   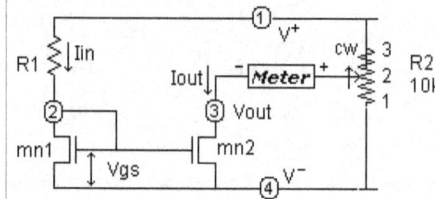

Turn the pot CCW so the node 3 volts equal the node 2 volts.

Measure the DC volts at nodes 1, 2, 3, 4.

Make measurements that produce an I_{out} plot of I_C(QN2) like Figure 90411 page 89.

Figure 906 Current Mirror Circuit 905

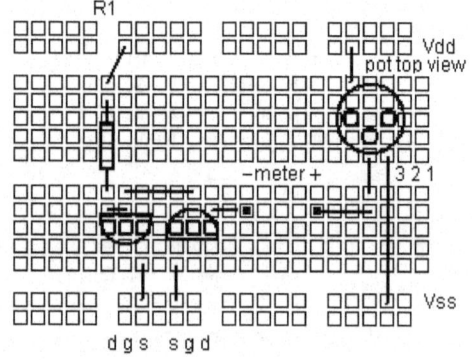

Calculate r_{out} from the slope of the I_{out} plot (Figure 90411).

Set the node 3 DC volts equal to the node 2 DC volts. Measure the currents at nodes 2 and 3. Are they equal? (Equation 4 page 88)

9.4 Differential Amplifier

The output voltage v_{out} is i_{out} times the output impedance r_{out} in parallel with any load impedance Z_L. The amplifier's r_{out} equals the output impedances of transistors mn_2 and mp_2 in parallel (Figure 907). At zero frequency the load impedance is infinite, because in an MOS circuit loads are capacitors. Design for gain A of 100 produces a value for W/L.

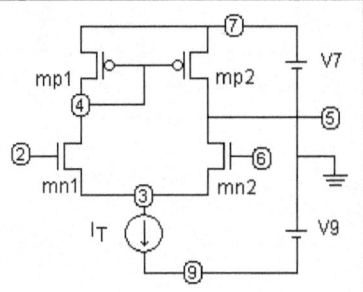

Figure 907 Differential Amplifier

(7a) $\quad i_{out} = i_{dsn2} + i_{dsp2} \quad$ and $\quad i_{dsp2} = i_{dsn1}$

(7b) $\quad i_{out} = i_{dsn2} + i_{dsn1} = g_{mn2}\dfrac{\Delta v}{2} + g_{mn1}\dfrac{\Delta v}{2} = (g_{mn2} + g_{mn1})\dfrac{\Delta v}{2} = g_{mn}\Delta v$

(8a) $\quad A = \dfrac{v_{out}}{v_2 - v_6} = \dfrac{r_{out} \cdot i_{out}}{\Delta v} = r_{out} \cdot i_{out} \cdot \dfrac{1}{\Delta v} = r_{out} \cdot g_m \Delta v \cdot \dfrac{1}{\Delta v} = g_m r_p \parallel r_n$

(8b) $\quad A = g_m \dfrac{\dfrac{1}{\lambda_p I_{DSp}}\dfrac{1}{\lambda_n I_{DSn}}}{\dfrac{1}{\lambda_p I_{DSp}} + \dfrac{1}{\lambda_n I_{DSn}}} = \dfrac{g_m}{I_{DS}} \cdot \dfrac{1}{(\lambda_p + \lambda_n)} = \dfrac{1}{(\lambda_p + \lambda_n)}\sqrt{\dfrac{2\mu C_{ox}}{I_{DS}}\dfrac{W}{L}}$

(8c) $\quad \dfrac{W}{L} = \left[A(\lambda_p + \lambda_n)\right]^2 \dfrac{I_{DS}}{2\mu C_{ox}} = \left[100(0.06+0.04)\dfrac{1}{V}\right]^2 \dfrac{30\mu A}{2 \cdot 180\mu A/V^2} = 8.33 \approx 10$

For W/L=7.5, 10, 15 Spice calculates gains 38.9, 39.8, 40.7 dB or 89, 97, 109 v_{out}/v_{in} (Figure 90732 page 92). The results show how subckts can expedite a design, that W/L is not critical, and that accuracy is not 100%.

We adapt transmission equation 20b page 98) where R replaces L.

(9) $\quad T_v = \dfrac{v_3}{v_1} \approx \dfrac{g_{m2}}{2y_L + g_{02}} = \dfrac{g_{m2}r_{02}}{1 + 2y_L} \qquad\qquad y_L = \dfrac{1}{R} + pC$

Where $R = \infty$, And $C = C_{BDmp2} + C_{BDmn2}$ (see Spice 9073 numeric data)

Electronic Circuits – Practical Learning

Figure 90731 Differential Amplifier DC transfer function

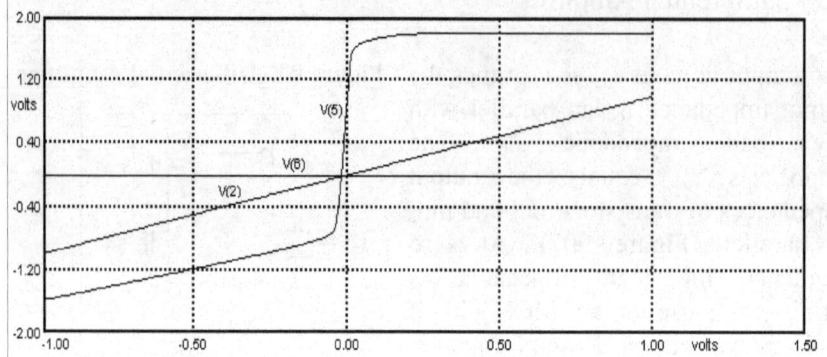

Gain is reduced by 1/2, because V_6 is a constant 0V, while V_2 varies from −1 to 1.

Figure 90732 Differential Amplifier AC response

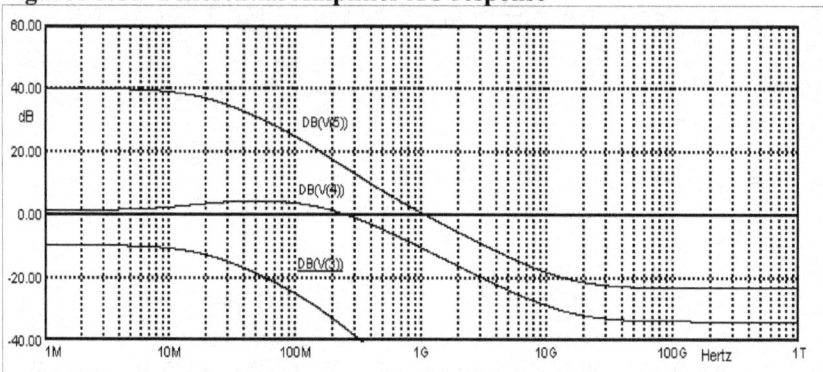

Gain = 100, because V_2 and V_6 are signal inputs.

Figure 90733 Differential Amplifier TRAN response

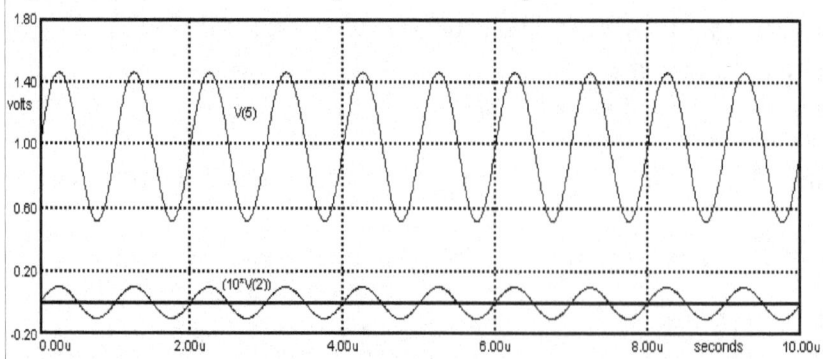

Spice program 9073

```
Fig9073.ckt   cmos diff amp
.include 180_N1P1.txt
.lib G:\!B\nlpbooks\ee103pdf\ec07spice\mos018_L.lib

V7 7 0 DC 1.8
V9 9 0 DC -1.8
V2 2 0 DC 0 AC 1
V6 0 6 DC 0 AC 1
R2 2 0 1E10
R6 6 0 1E10
xmp1 4 4 7 7 mp5L        ; body connected to +1.8volts
xmp2 5 4 7 7 mp5L        ; W/L=10 all xmn and xmp
xmn1 4 2 3 9 mn5L        ; body connected to -1.8volts
xmn2 5 6 3 9 mn5L
IT 3 9 DC 60u
.AC DEC 1000 1e+006 1e+012
.PLOT AC VDB(3) VDB(5) -40,60
.PRINT AC V(5)
.PRINT AC VDB(3) VDB(5)

.DC V2 -1 1 0.01
.PLOT DC V(2) V(6) V(5) -2,2
.PRINT DC V(2) V(5)

.TRAN 1e-008 1e-005 0 1e-008
.PLOT TRAN (10*V(2)) V(5) -0.5,1
.TEMP 27
.end
```

```
Numeric data
           XMP1.MP5L      XMP2.MP5L      XMN1.MN5L      XMN2.MN5L
Model   P1             P1             N1             N1
CBD     24.923f        24.923f        17.758f        17.758f
CBS     60.839f        60.839f        35.524f        35.524f
```

The bodies of mn_1 and mn_2 are connected to $-1.8V$.

WARNING In a $\pm V$ volt system do not connect nmos bodies to 0v. If they are connected to 0V, then, because their sources are at a negative voltage, the body to channel parasitic pn diode would be turned on ($V_{PN}>0$) and diode current would be added to the source current. These undesirable currents show up in the Spice numeric data, which shows all MOS currents.)

Electronic Circuits – Practical Learning

Practice: Build and Test the Differential Amplifier (Figure 908) on the breadboard. $R_1 = 12K$ is sized for $I_1 = 0.8mA$. The layout is not easy to do. See Figure 909. Use $\pm5V$ supplies, 2N7000 NMOS transistors, and ZVP3306a PMOS transistors.

Connect the function generator output to node 2 via $C_1 \geq 1\mu F$. Adjust the pot so that $V_4 = V_5$.

Figure 908 Differential Amplifier in a circuit

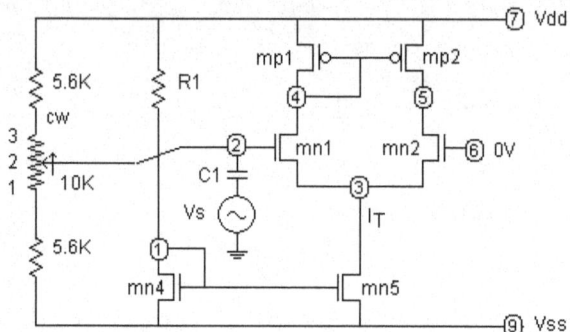

Figure 909 Differential Amplifier in a circuit

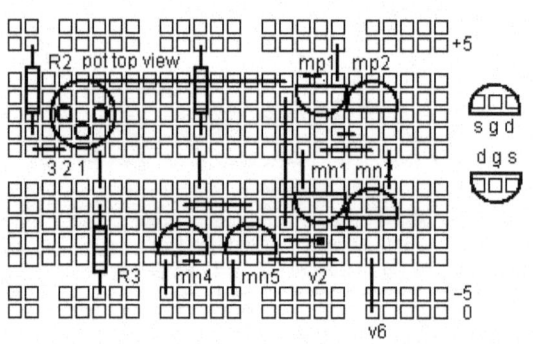

Steady State Response Select sine wave output. Set frequency to 1KHz (*1000μs period*). Connect oscilloscope channel 1 to node 2, and channel 2 to node 5. Monitor node 5. *Set oscilloscope channel 2 to 2V/cm.* Set 0V level at center screen. Select time base to be *200μs/cm so that full screen 10 cm sweep takes 2000μs.* Adjust the V_S signal level so that V_5 is an undistorted sine wave spanning 5cm. This level is 0dB.

Measure the gain V_5/V_2 (about 38).
Measure the frequency response −1dB (5.68KHz), −3dB (11.36KHz).
Reminder: these transistors are different from the N1 and P1 models.

96

Process Parameters ($L_{min} = 2\lambda = 0.18 \mu m$, $t_{ox} = 4.08nm$)

$$C_{ox} = \frac{\varepsilon_{SiO2}\varepsilon_0}{t_{ox}} = \frac{3.9 \times 8.85}{0.004} \frac{aF}{\mu m \times \mu m} = 8629 \frac{aF}{\mu m^2} = 8.63 \frac{fF}{\mu m^2}$$

$$C_{gate} = C_{ox}WL = C_{ox}\frac{W}{L}L^2 = C_{ox}L^2\frac{W}{L}$$

$$= 8.63\frac{fF}{\mu m^2}(0.18\mu m)^2\frac{W}{L_{min}} = 0.280\frac{W}{L_{min}}fF$$

Transistor drive current (C in fF)

nmos i_{drive}=0.60 mA/μm   pmos i_{drive}=0.26 mA/μm

$$\frac{W}{L} = \frac{1}{L}\frac{C_{load}}{i_{drive}}\frac{dv}{dt}$$

$$\frac{W_p}{L_{min}} = \frac{1}{0.18\mu m} \cdot \frac{C_{load}\ fF}{0.26mA/\mu m} \cdot \frac{1.8V}{100pS} = 0.3846C_{load}\ fF$$

$$\frac{W_n}{L_{min}} = \frac{1}{0.18\mu m} \cdot \frac{C_{load}\ fF}{0.60mA/\mu m} \cdot \frac{1.8V}{100pS} = 0.1667C_{load}\ fF$$

Resistance

$R_{metal\ k} = 0.078\ \Omega/sq$ (k = 1, 2, 3, 4, 5, 6) $R_{poly} = 7.8\ \Omega/sq$

Capacitance

Layer	aF/μm	line width	line sep μm	aF/λ	width λ
poly	144	0.16	0.27	13.0	2
metal 1	244	0.23	0.23	26.0	3
	110	0.23	2.00	11.6	3
metal 2	220	0.28	0.28	19.1	3
	85	0.28	2.00	7.4	3
metal 3,4,5	215	0.28	0.28	18.7	3
	78	0.28	2.00	6.8	3
metal 6	260	0.44	0.46	19.2	4
	110	0.44	2.00	8.1	4

9.5 LC Tuned Amplifier

Electronic communications hardware is implemented by use of *frequency selective* circuits. Frequency selective circuits define *frequency bands* that confine the communications signals to those defined bands so that *communication channels* do not interfere with each other. Resonant circuits, *LC tuned circuits,* implement the frequency selective concept.

7.5.1 LC Tuned Amplifier Design

A tuned amplifier designed for use in an integrated circuit is a differential amplifier biased by a current mirror (Figure 910). The drain currents (Figure 91011) show that the transfer function is linear. Parasitic capacitance is found by plotting the LC frequency response with capacitor C_1 removed from the circuit. L_1=100µH. The parasitic resonant frequency is about 250MHz (Figure

Figure 910 Tuned Amplifier

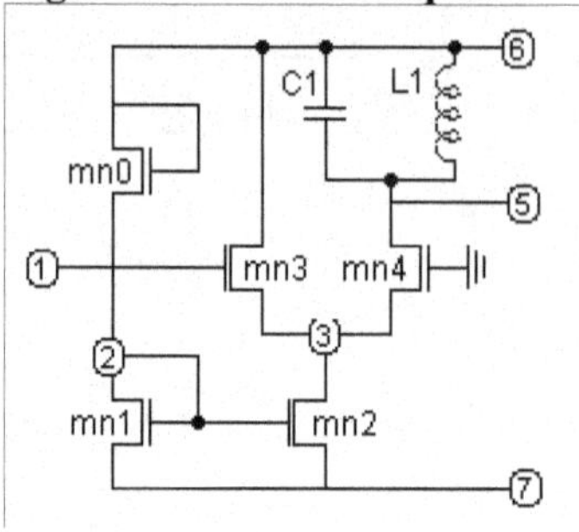

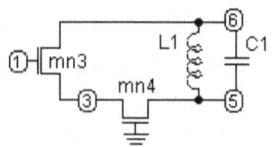

91012), and C_P=0.004pF. Adding C_1=100pF changes the resonant frequency to 1.591MHz (Figure 91013 page 101). Gain at resonance is 38dB (Figure 91012).

$$(10) \quad C_P = \frac{1}{\omega_0^2 L} = \frac{1}{4\pi^2 (250 \cdot 10^6)^2 \cdot 100 \cdot 10^{-6}} = 4fF$$

$$(11) \quad f_0 = \frac{1}{2\pi\sqrt{L(C_P + C_1)}} = \frac{1}{2\pi\sqrt{100 \cdot 10^{-6} \times 100.004 \cdot 10^{-12}}} = 1.591MHz$$

The amplifier's AC schematic reduces to the mn_3 emitter follower driving common gate amplifier mn_4 (Figure 911). mn_4's collector drives the LC tank circuit. MOS parasitic capacitors are omitted from the AC small signal equivalent circuit, because the poles they create are at significantly higher frequencies than the resonant frequency. The transmission function T_v is derived from the AC equivalent circuit (Figure 912).

Figure 911 AC Schematic

Figure 912 AC Equivalent Circuit

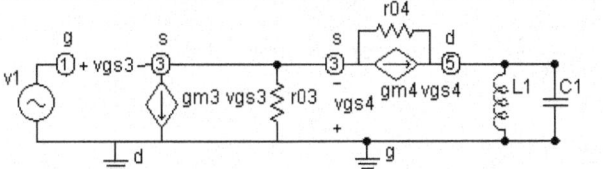

This analysis assumes r_{03} and r_{04} are infinite.

(12a) $\quad 0 = g_{m4}v_{gs4} + y_L v_3 \qquad$ node 3

(12b) $\quad 0 = -g_{m3}v_{gs3} - g_{m4}v_{gs4} \qquad$ node 2

(12c) $\quad v_{gs4} = -v_2 \qquad\qquad v_{gs3} = v_1 - v_2$

(13a) $\quad 0 = -g_{m4}v_2 + y_L v_3 \qquad\qquad$ node 3

(13b) $\quad 0 = -g_{m3}(v_1 - v_2) - g_{m4}(-v_2) \quad$ node 2

(13c) $\quad 0 = -g_{m3}v_1 + (g_{m3} + g_{m4})v_2 \qquad$ node 2

(14) $\quad T_v = \dfrac{v_3}{v_1} = \dfrac{v_3}{v_2} \times \dfrac{v_2}{v_1} = \dfrac{g_{m4}}{y_L} \times \dfrac{g_{m3}}{g_{m3} + g_{m4}} = \dfrac{g_{m3}g_{m4}}{g_{m3} + g_{m4}} z_L$

Analysis when r_{03} and r_{04} are included in the analysis.

(15a) $\quad 0 = g_{m4}v_{gs4} + (g_{04} + y_L)v_3 - g_{04}v_2 \qquad\qquad$ node 3

(15b) $\quad 0 = -g_{m3}v_{gs3} - g_{m4}v_{gs4} + (g_{03} + g_{04})v_2 - g_{04}v_3 \quad$ node 2

(15c) $\quad v_{gs4} = -v_2 \qquad\qquad v_{gs3} = v_1 - v_2$

(16a) $\quad 0 = -(g_{m4} + g_{04})v_2 + (g_{04} + y_L)v_3 \qquad\qquad$ node 3

(16b) $\quad 0 = -g_{m3}v_1 + (g_{03} + g_{04} + g_{m3} + g_{m4})v_2 - g_{04}v_3 \quad$ node 2

(17) $\quad g_{m3}v_1 = (g_{03} + g_{04} + g_{m3} + g_{m4}) \dfrac{(g_{04} + y_L)v_3}{(g_{m4} + g_{04})} - g_{04}v_3$

$\qquad g_{m3}\dfrac{v_1}{v_3} = \dfrac{(g_{03} + g_{04} + g_{m3} + g_{m4})y_L + (g_{03} + g_{04} + g_{m3} + g_{m4})g_{04} - (g_{m4} + g_{04})g_{04}}{(g_{m4} + g_{04})}$

$\qquad g_{m3}\dfrac{v_1}{v_3} = \dfrac{(g_{03} + g_{04} + g_{m3} + g_{m4})y_L + (g_{03} + g_{m3})g_{04}}{(g_{m4} + g_{04})}$

(18) $T_v = \dfrac{v_3}{v_1} = \dfrac{g_{m3}(g_{m4} + g_{04})}{(g_{03} + g_{04} + g_{m3} + g_{m4})y_L + (g_{03} + g_{m3})g_{04}}$

check $T_v(g_0 = 0) = \dfrac{g_{m3}(g_{m4})}{(g_{m3} + g_{m4})y_L}$ compare to equation 39

Numeric output data shows $g_{m3} = g_{m4} \gg g_0$.

(19) $T_v = \dfrac{v_3}{v_1} \approx \dfrac{g_{m3}g_{m4}}{(g_{m3} + g_{m4})y_L + (g_{m3})g_{04}} = \dfrac{g_{m4}}{2y_L + g_{04}} = \dfrac{1}{2}\dfrac{g_{m4}}{y_L + 0.5g_{04}}$

Gain at resonance is 38.1dB (Figure 91012 page 101 shows 38dB).
Note: $y_L = 0$ at resonance.

(20a) $g_{m3} = g_{m4} = 0.630\text{mA}/\text{V}$ $r_{03} = r_{04} = 1/G_{DS} = 10^6/7.8 = 128K$

(20b) $T_v(\omega_0) = \dfrac{v_3}{v_1} = = \dfrac{1}{2}\dfrac{g_{m4}}{y_L + 0.5g_{04}} = \dfrac{0.630}{2}\dfrac{\text{ma}}{\text{V}} \cdot 2 \cdot 128K = 80.6 = 38.1\text{dB}$

Q_P (two ways) and Bandwidth BW

(21a) $Q_p = \dfrac{2r_{04}}{\omega_0 L_1} = \dfrac{256 \cdot 10^3}{2\pi \cdot 1.591 \cdot 10^6 \times 100 \cdot 10^{-6}} = 256$

(21b) $Q_p = \dfrac{2r_{04}}{\omega_0 L_1} = \dfrac{2r_{04}\sqrt{L_1 C_1}}{L_1} = \dfrac{2r_{04}\sqrt{C_1}}{\sqrt{L_1}} = \dfrac{2r_{04}}{z_0} = \dfrac{256K}{1K} = 256$

(22) 3 dB BW $= \dfrac{f_0}{Q_p} = \dfrac{1.591 \cdot 10^6}{256} = 6.2$ KHz

MHz	v1	v5 dB (Spice 9101 data)	
1.588	0.000	35.047	
1.589	0.000	36.075	
1.590	0.000	36.981	
1.591	0.000	37.668	
1.591	0.000	37.986	compare to eqn44b & eqn 36
1.592	0.000	37.877	
1.593	0.000	37.292	
1.597	0.000	36.756	
1.597	0.000	35.783	

Figure 91011 Tuned amplifier Drain currents

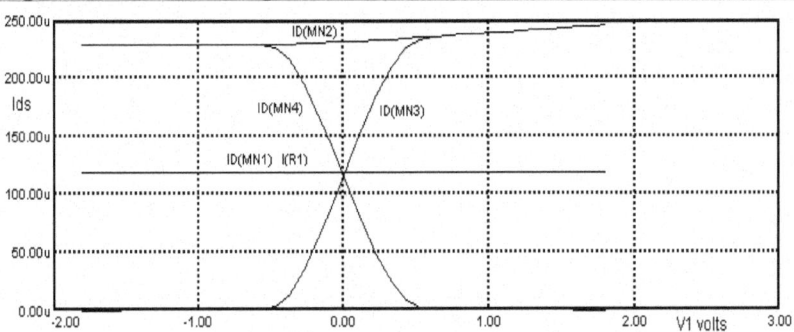

Figure 91012 Tuned amplifier frequency response $C_1=0$

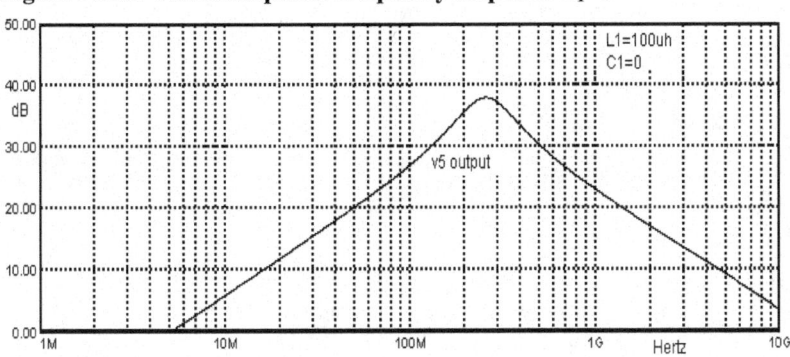

Figure 91013 Tuned amplifier frequency response $C_1=100pF$

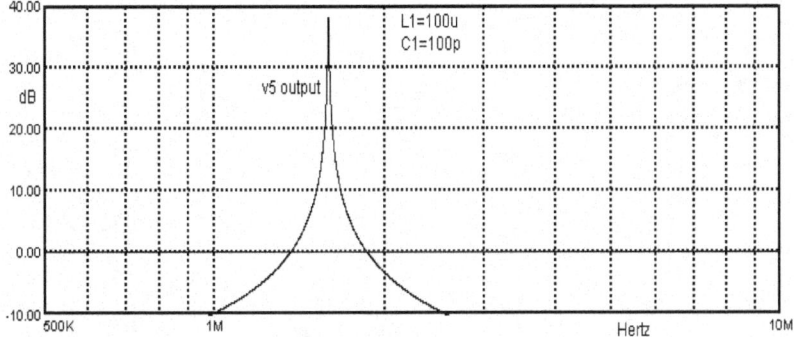

Spice program 9101

```
Fig9101.ckt   nmos tuned amplifier
.include 180_N1P1.txt
V1 1 0 DC 0 AC 1 sin(0 0.1 1.5715E6 0 0)
Vdd 6 0 DC 1.8
Vss 7 0 DC -1.8


mn3 6 1 3 7 N1 L=0.9u W=9.00u
mn4 5 0 3 7 N1 L=0.9u W=9.00u

*R1 6 2 20K      ;115uA R1 replaced by mn0
mn0 6 6 2 7 N1 L=0.9u W=0.90u
mn1 2 2 7 7 N1 L=0.9u W=1.80u    ; W=4.5u
mn2 3 2 7 7 N1 L=0.9u W=3.60u


*R2 6 5 250K
L1 6 5 100u
C1 6 5 100p

.TRAN 1e-008 2e-005 0 1e-009
.PLOT TRAN V(1) V(3) V(5) -1,5

*.PLOT DC V(1) V(3) V(5) -2,2

*.PLOT AC VDB(1) VDB(5) -50,50

.DC V1 -1.8 1.8 0.1
.PLOT DC ID(MN1) ID(MN2) ID(MN3) ID(MN4) 0,250U
.PRINT DC ID(MN1) ID(MN2) ID(MN3) ID(MN4)
.AC DEC 5020 500000 1e+007
*.AC DEC 5020 1e+006 1e+010
.TEMP 27
.PLOT AC VDB(1) VDB(5) -10,40
.PRINT AC VDB(1) VDB(5)
.end
```

MOSFET Devices

	MN3	MN4	MN0	MN1	MN2
Model	N1	N1	N1	N1	N1
ID	18.455u	118.455u	121.033u	121.033u	236.910u
VGS	964.456m	964.456m	2.345	1.255	1.255
VDS	2.764	2.764	2.345	1.255	835.544m
VBS	-835.544m	-835.544m	-1.255	0.000	0.000
VTH	665.950m	665.950m	754.531m	51.211m	451.211m
VDSAT	268.452m	268.452m	1.206	611.832m	611.832m
GM	637.082u	637.082u	88.618u	224.112u	440.014u
GDS	7.959u	7.959u	2.926u	5.658u	15.028u
GMB	166.795u	166.795u	40.640u	89.689u	175.766u

Practice: Build and Test the Tuned Amplifier. (Figure 913) Use ±5V supplies, and 2N7000 n channel transistors, L_1=1000μH and C_1=2700pF. Since mn_0 is not available, replace it with R_0 sized for I_1=0.8mA. Calculate R_0 (12K). Install R_0 and measure I_1.

Connect the function generator output to node 1 via C_1>=1μF. Adjust the pot so that V_5 equals about 2.2V DC.

The *reactance chart* shows that the LC resonant frequency (96.8KHz) and characteristic impedance (608 ohms) values are vertical and horizontal lines passing through the point where the L_1=1000μH, and C_1=2700pF reactance lines cross.

Calculate resonant frequency f_0. Calculate the characteristic impedance z_0. Compare your numbers to the reactance chart numbers.

Figure 913 Tuned Amplifier in a circuit

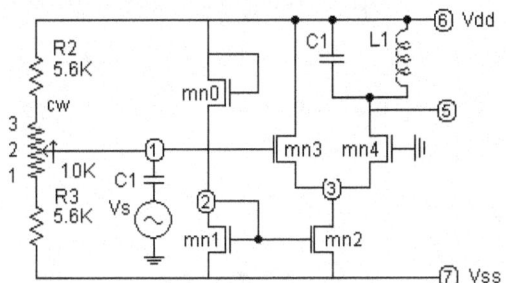

Steady State Response Connect the function generator's output to node 1. Select a 100KHz sine wave. Connect oscilloscope channel 1 to node 5. *Set oscilloscope channel 1 to 2V/cm.*

Set 0V level at center screen. Select time base to be *5μs/cm so that full screen 10 cm sweep takes 50μs.*

Adjust signal generator voltage so that the signal spans 5cm on the oscilloscope screen (0dB) and is an undistorted sine wave. You may have to reduce signal generator level as the frequency approaches resonance.

Adjust the frequency slowly until you find a maximum at resonant frequency f_0 (104KHz). Compare f_0 to 96.8KHz. At what frequencies above and below f_0 does v_5/v_1=0.707? (−3dB) What is the bandwidth? (5KHz) Q? (104/5) (equation 22 page 100)

Measure the gain at f_0. Calculate the gain (equation 20b page 98) and compare to your measurement. Work out the gain and Q differences.

Figure 914 Tuned Amplifier in a circuit

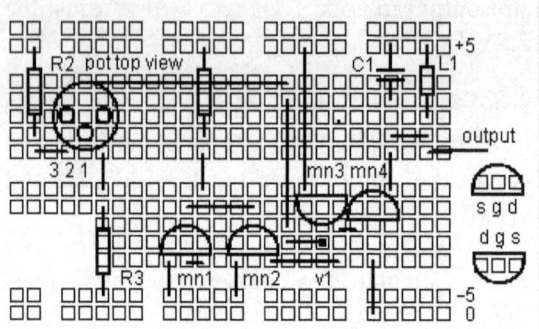

Compare f_0 to 96.8KHz. At what frequencies above and below f_0 does $V_2/V_1=0.707$? (−3dB, 106KHz, 101KHz) What is the bandwidth? (5KHz) Q? (104/5)

Measure the gain at f_0. Calculate the gain and compare to your measurement. Work out the gain and Q differences.

Transient Response of the circuit.

On the Function Generator switch from sine wave to square wave and do not change frequency. What do you see at node 5? What did you expect to see?

9.6 LC Oscillator

An LC oscillator is a tuned amplifier with feedback. The $L_1C_1L_2$ circuit is a high pass filter (Figure 915). At the oscillation frequency there is a 180° phase shift through this filter, and transmission of a signal from transistor mn_4's base to collector produces another 180° phase shift. The 360° shift satisfies one requirement for oscillation. The other requirement for oscillation is a return ratio

Figure 915 Hartley Oscillator

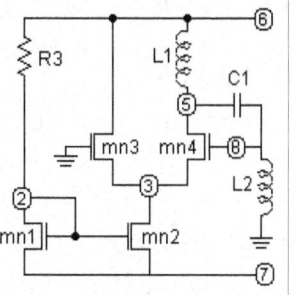

>1. Disconnect node 8 from mn_4's gate. The return ratio is the transmission from mn_4's gate to node 5 to node 8. The upcoming circuit analysis is confirmed by Spice Figure 91521 page 107.

Figure 916 Hartley Model

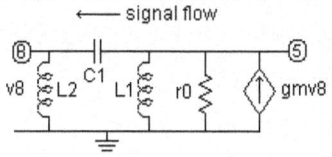

Write the KCL node equations.

$$(23) \quad 0 = g_m v_8 - pC_1 v_8 + \left(g_0 + pC_1 + \frac{1}{pL_1} \right) v_5$$

$$(24) \quad 0 = \left(pC_1 + \frac{1}{pL_2} \right) v_8 - pC_1 v_5$$

Solve 24b for v_5.

$$(25) \quad v_5 = \left(1 + \frac{1}{p^2 L_2 C_1} \right) v_8$$

Substitute 25 into 23.

$$(26a) \quad 0 = g_m v_8 - pC_1 v_8 + \left(g_0 + pC_1 + \frac{1}{pL_1} \right) \left(1 + \frac{1}{p^2 L_2 C_1} \right) v_8$$

$$(26b) \quad 0 = g_m + \left(g_0 + \frac{1}{pL_1} \right) + \left(g_0 + \frac{1}{pL_1} + pC_1 \right) \left(\frac{1}{p^2 L_2 C_1} \right)$$

(26c) $\quad 0 = g_m + \left(g_0 + \dfrac{1}{pL_1}\right) \times \left(1 + \dfrac{1}{p^2 L_2 C_1}\right) + pC_1 \left(\dfrac{1}{p^2 L_2 C_1}\right)$

(26d) $\quad 0 = g_m + \dfrac{1}{pL_1}(pL_1 g_0 + 1) \times \dfrac{1}{p^2 L_2 C_1}\left(1 + p^2 L_2 C_1\right) + \dfrac{1}{pL_2}$

(26e) $\quad 0 = g_m\left(p^3 L_1 L_2 C_1\right) + \left(1 + g_0 pL_1\right)\left(1 + p^2 L_2 C_1\right) + p^2 L_1 C_1$

$\qquad 0 = \left(1 + pL_1 g_0\right) + p^2 L_1 C_1 + \left(1 + pL_1 g_0\right)\left(p^2 L_2 C_1\right) + p^3 L_1 L_2 C_1 g_m$

$\qquad 0 = 1 + pL_1 g_0 + p^2 L_1 C_1 + p^2 L_2 C_1 + p^3 L_1 L_2 C_1 g_0 + p^3 L_1 L_2 C_1 g_m$

(27) $\quad$ let $p = j\omega$

$0 = 1 + j\omega L_1 g_0 - \omega^2 L_1 C_1 - \omega^2 L_2 C_1 - j\omega^3 L_1 L_2 C_1 g_0 - j\omega^3 L_1 L_2 C_1 g_m$

$0 = \left[1 - \omega^2(L_1 + L_2)C_1\right] + j\omega\left[L_1 g_0 - \omega^2 L_1 L_2 C_1(g_0 + g_m)\right]$

Real part equals zero produces resonant frequency.

(28a) $\quad 0 = 1 - \omega_0^2(L_1 + L_2)C_1 \quad \Rightarrow \quad \omega_0^2 = \dfrac{1}{(L_1 + L_2)C_1}$

Imaginary part equals zero produces gain criteria.

(28b) $\quad 0 = j\omega_0\left[L_1 g_0 - \omega_0^2 L_1 L_2 C_1(g_0 + g_m)\right]$

$\Rightarrow \omega_0^2 = \dfrac{L_1 g_0}{L_1 L_2 C_1(g_0 + g_m)} = \dfrac{g_0}{L_2 C_1(g_0 + g_m)} = \dfrac{1}{L_2 C_1(1 + g_m r_0)}$

(28c) $\quad$ let $5la = 5lb\,\dfrac{1}{(L_1 + L_2)C_1} = \dfrac{1}{L_2 C_1(1 + g_m r_0)} \quad \Rightarrow \quad g_m r_0 = \dfrac{(L_1 + L_2)}{L_2} - 1$

Numerical values

(29a) $\quad \omega_0 = \left(\dfrac{1}{(L_1 + L_2)C_1}\right)^{0.5} = \left(\dfrac{1}{112\mu H \cdot 56 pF}\right)^{0.5} = 12.627 \cdot 10^6$

$\qquad f_0 = \dfrac{\omega_0}{2\pi} = \dfrac{12.627}{2\pi} = 2.0096 MHz$

(29b) $\quad g_m r_0 \geq \dfrac{(L_1 + L_2)}{L_2} - 1 = \dfrac{112\mu H}{56\mu H} - 1 = 2 - 1 = 1$

Figure 91521 MOS Hartley LC Oscillator

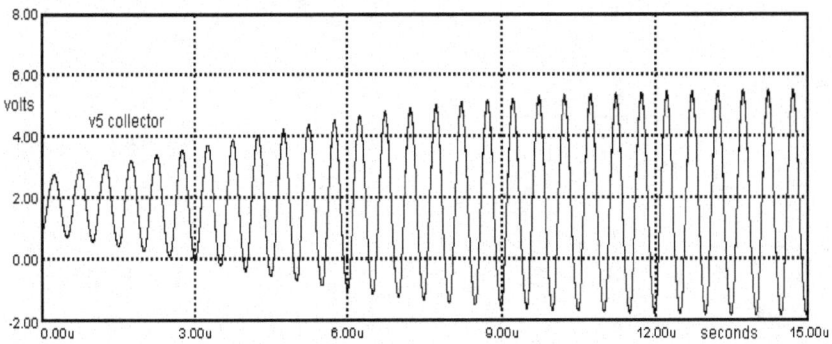

Equation 52a predicts f_0=2.0096MHz. Figure 91521 shows period T=3µs/6 cycles=0.5µs so that frequency = 1/T = 2MHz.

Spice Program 9152 Oscillator

```
Fig9152.ckt  nmos tuned amplifier as oscillator

.include 180_N1P1.txt

Vdd 6 0 DC 1.8
Vss 7 0 DC -1.8

R1 6 2 20K
mn1 2 2 7 7 N1 L=0.9u W=1.80u
mn2 3 2 7 7 N1 L=0.9u W=3.60u

mn3 6 0 3 7 N1 L=0.9u W=3.60u
mn4 5 8 3 7 N1 L=0.9u W=3.60u

L1 6 5 56u
C1 5 8 56p
L2 8 0 56u

.TRAN 1e-012 1.5e-005 0 1e-009 UIC
* NOTE UIC disables DC operating point calculation which
* may take forever
.TEMP 27
.PLOT TRAN V(5) -2,8
.end
```

Practice: Build and Test the Hartley Oscillator. (Figure 711) Use ±5V supplies, and 2N7000 n channel transistors. Calculate a value for R_3 so that I_1=0.8mA (12K).

Figure 915

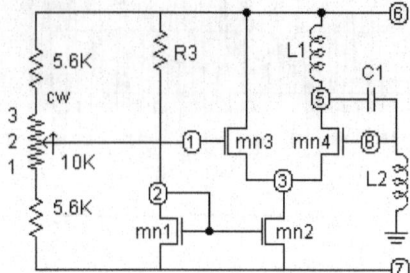

Figure 917

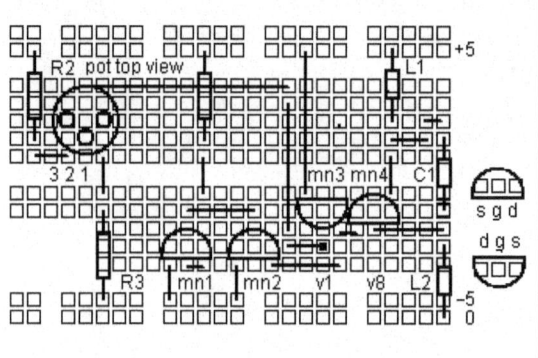

L_1=1000µH C_1=680pF L_2=1000µH

Adjust the pot so that V_5 is about 2.2V, and the node 5 signal is essentially undistorted.

Oscillates at 165KHz - about 3V peak to peak. (plus third harmonic, which is about 0.3V peak to peak)-

Calculate the theoretical resonant frequency (equation 29a page 106).

9.7 CMOS Digital Circuit Design

This is about circuit design of two CMOS digital circuits. This is *not* about logic design. The pmos and nmos transistors are referred to as complements in a CMOS digital circuit. If pmos is on, then nmos is off, and vice versa. Pmos on raises a node's voltage to V_{DD} (such as 1.8V), and nmos on lowers the node voltage to V_{SS} (0V). All circuits are in fact analog circuits that are usually distinguished by *linear* input to output voltage or current transfer functions. A circuit is 'digital' simply because, by design, input node voltages are either at V_{SS} (L) or V_{DD} (H).

A two-level H and L voltage system is designed to represent switching algebra values 1 and 0. The assignment of values 0 and 1 to node voltage levels can be made in three ways that results in logic systems known as positive logic (at all nodes H=1, L=0), negative logic (at all nodes H=0, L=1), and mixed logic (some nodes positive and some nodes negative).

There are two radically different classes of digital CMOS circuits. One is based on the transmission gate[4] (which we do not discuss here), and the other is based on the static CMOS inverter.

Static Logic Circuits Static logic gates are assemblies of *merged* inverters where you find transistors in series and in parallel. A circuit is referred to as *static* when each circuit node is connected to a voltage source by a turned on transistor, when all circuit inputs are either H or L. Static because a turned on transistor compensates for any leakage currents draining a node, thereby maintaining the charge on the node capacitance so that the voltage from the node to ground does not change.

The basic Inverter and a 2-input NAND gate are designed.

Circuit Loads Any CMOS circuit output is connected via metal and polysilicon lines to input gates of driven circuits. These physical lines and MOS gates have capacitance to ground. We name the value of this capacitance C_{LOAD}.

Design The design question is *How do we calculate the transistors W/L?*

---

[4] N. L. Pappas, CMOS Circuit Design - Analog, Digital, IC Layout

9.8 CMOS Inverter

The basic CMOS circuit is the inverter (Figure 918) where one nmos device is paired with one pmos device. When the nmos is on it discharges the output node to 0V, and when the pmos is on it charges the output node to 1.8V in a 1.8V system. Voltage transitions from 0 to 1.8V or 1.8V to 0V are referred to as rail to rail transitions. With rail to rail transitions either nmos is on or pmos is on. This is what is needed for 2-level H and L digital logic. *CMOS requires each nmos to be paired with a pmos.*

Figure 918 CMOS inverter   **Figure 919 Inverter driving C_{load}**

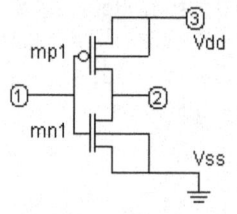
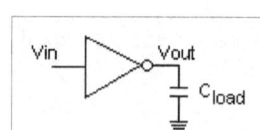

In a CMOS digital circuit C_{LOAD} (Figure 919) draws a *current* i_{LOAD} *only* while the applied voltage switches from L to H or H to L. For example:

$$(30) \quad i_{load} = C_{load}\frac{dv}{dt} \quad \Rightarrow \quad i_{load} = 100fF\frac{1.8V}{200ps} = 0.9mA$$

Any MOS semiconductor process defines nmos and pmos drive current capability. The drive current process parameters for the short channel $L_{MIN}=2\lambda=0.18\mu m$ process (page 97) are

$$(31a) \quad i_{ndrive} = 0.60\frac{mA}{\mu m} \quad (31b) \quad i_{pdrive} = 0.26\frac{mA}{\mu m} \quad (31c) \quad i_{ds} = i_{drive}\frac{mA}{\mu m}\times W\,\mu m$$

The drain current I_{ds} that is proportional to W, must be greater than or equal to i_{LOAD}.

$$(32) \quad W = \frac{i_{ds}}{i_{drive}} = \frac{1}{i_{drive}}i_{load} = \frac{1}{i_{drive}}C_{load}\frac{dv}{dt}$$

In a 1.8V system, when $C_{LOAD} = 100fF$, and switching time is 200pS, the inverter pmos and nmos transistors are sized as follows.

$$(33a) \quad \frac{W_n}{L} = \frac{1}{L}\frac{C_{load}}{i_{ndrive}}\frac{dv}{dt} = \frac{1}{0.18\mu m}\cdot\frac{100fF}{0.6mA/\mu m}\cdot\frac{1.8}{200ps} = \frac{1.50\mu m}{0.18\mu m} = 8.33 \rightarrow \frac{16.7\lambda}{2\lambda}$$

$$(33b) \quad \frac{W_p}{L} = \frac{i_{ndrive}}{i_{pdrive}}\frac{W_n}{L} = \frac{0.60mA}{0.26mA}\cdot\frac{17\lambda}{2\lambda} = \frac{39.2\lambda}{2\lambda} \rightarrow \frac{40\lambda}{2\lambda}$$

The ideal V_4 waveform shows that output voltage V_2 transient response waveform confirms these calculations (Figure 71311).

Figure 91811 CMOS inverter with Cload 100fF

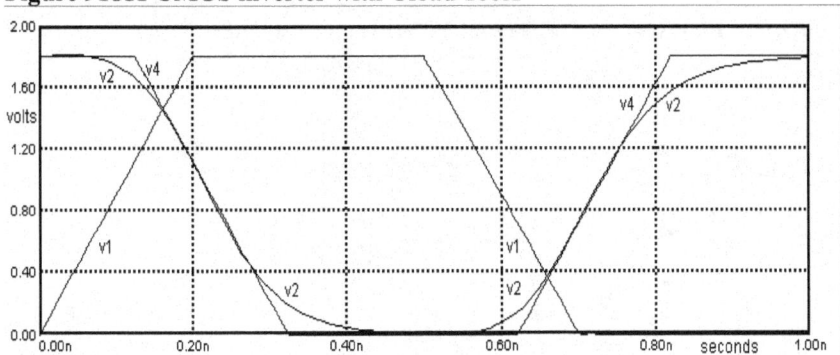

Spice program 91811

```
Fig9181.ckt  cmos inverter
Vdd 3 0 DC 1.8

V1 1 0 PULSE(0    1.8 000p 200p 200p 300p 1000p)
V4 4 0 PULSE(1.8     0 125p 200p 200p 285p 1000p)

* Level 8 (BSIM) SPICE models
.include 180_N1P1.txt

MP1 2 1 3 3 P1 L=0.18u  W=3.6u    ;W/L=3.6/0.18=40/2
MN1 2 1 0 0 N1 L=0.18u  W=1.5u    ;W/L=1.5/0.18=16.7/2
C2 2 0 100f

.TRAN 1e-011 1e-009 0 1e-011
.TEMP 27
.PLOT TRAN V(1) V(2) V(4) 0,2
.end
```

Electronic Circuits – Practical Learning

Practice: Build and Test the CMOS Inverter

Select $V_{DD} = 5V$, $V_{SS} = 0V$, 2N7000 NMOS transistor, and ZVP3306a PMOS transistor.

Transient Response of the circuit.
Connect the Function Generator square wave output to node 1. Set the frequency to 10KHz. What do you see at node 2? What did you expect to see?

Figure 918 Inverter   **Figure 925 Inverter Layout**

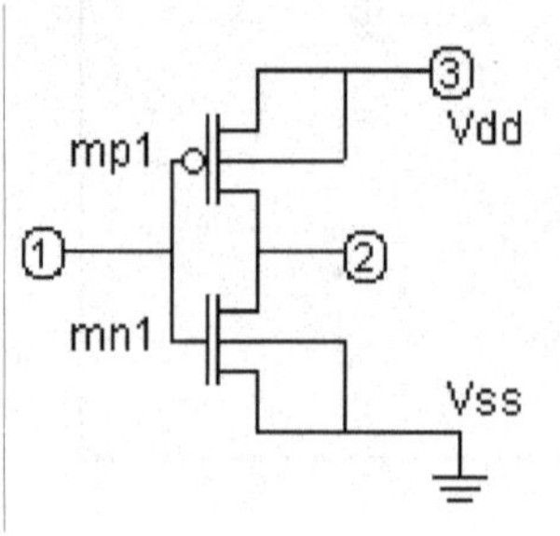

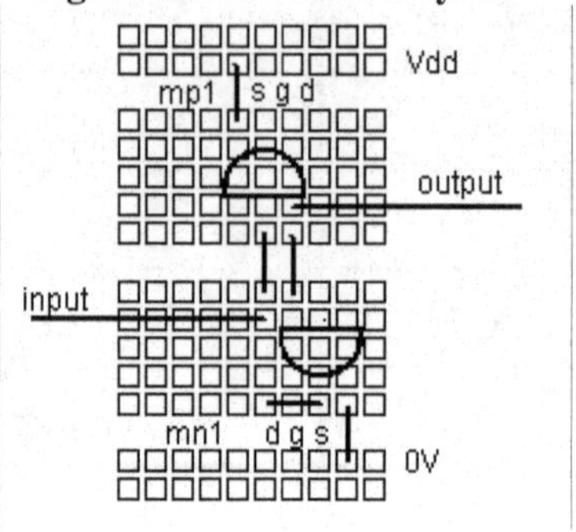

9.8 CMOS NAND

The AND condition is created when two nmos mn_1, mn_2 (Figure 920) are in series and both are turned on by H levels at their gates discharging output node 25 to L. Each nmos has its pmos partner that has to charge the output to H when its nmos is off. This is why the two pmos mp_1, mp_2 are in parallel, because *either* has to charge the output to H when turned on by L levels at a gate. The NAND gate is based on two cleverly merged inverters.

Figure 920 CMOS NAND

Fanout In a cmos logic system fanout F is defined as a ratio of capacitors C_{load}/C_{in}. Each capacitor can be expressed in terms of MOS parameters and system switching time Δt required to traverse system rail-to-rail voltage Δv. The resulting equation for F can be used to design the system.

$$(34) \quad i = C\frac{dv}{dt} \qquad C_{in} = W_{in}LC_{ox} \qquad C_L = I_L\frac{\Delta t}{\Delta V}$$

$$(35) \quad F = \frac{C_L}{C_{in}} = \frac{1}{C_{ox}} \cdot \frac{1}{L} \cdot \frac{\Delta t}{\Delta V} \cdot \frac{I_L}{W_{in}}$$

The inverter was designed to drive 100fF (9.8). The NAND has two nmos in series. One can demonstrate that two nmos in series, each with width W, have an equivalent width of W/2. Double W to $W_N = 2\times16.7\lambda = 34\lambda$ when the inverter numbers are used (equations 33).

$$(36) \quad C_{in} = W_{in}LC_{ox} = (W_p + W_n)LC_{ox}$$

$$= (40+34)\lambda\frac{0.09\mu m}{\lambda} \cdot 0.18\mu m\frac{8.63fF}{\mu m^2} = 10.35fF$$

This circuit can drive 10 circuits, because the fanout F=100/10.35 = 9.66 $\approx$ 10. The ideal V_4 waveform shows that output voltage V_{25} transient response confirms these calculations (Figure 92011).

Figure 92011 2-input CMOS NAND gate Transfer function V25/V1

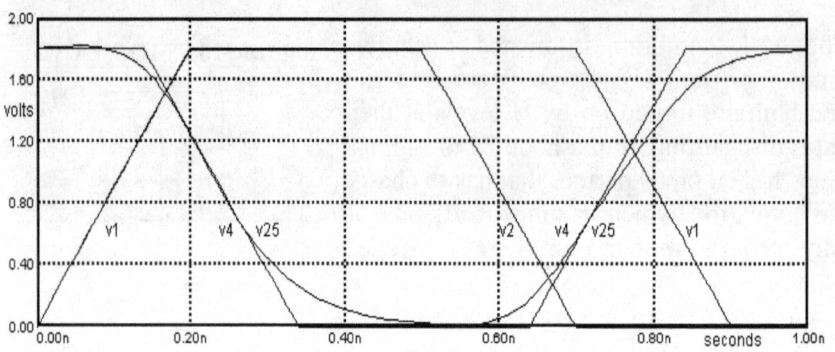

Spice program 9201

```
Fig9201.ckt   2-in NAND Transient Response
.include 180_N1P1.txt
.subckt mn1 103 102 101
MN1 103 102 101 0 N1 L=0.18u W=0.36u
+ AD=0.1782p AS=0.1782p PD=1.71u PS=1.71u    ;4/2=W/L
.ends mn1

.subckt mn6 118 117 116
MN6 118 117 116 0 N1 L=0.18u W=2.70u
+ AD=1.3365p AS=1.3365p PD=6.39u PS=6.39u    ;30/2=W/L
.ends mn6

.subckt mp7 221 220 219 298
MP7 221 220 219 298 P1 L=0.18u W=3.60u
+ AD=1.7820p AS=1.7820p PD=8.19u PS=8.19u    ;40/2=W/L
.ends mp7
XMP1   25 1 99 99 mp7      ;40/2
XMP2   25 2 99 99 mp7      ;40/2
XMN11 25 1 12    mn6     ;30/2
XMN12 25 1 12    mn1     ; 4/2
XMN21 12 2  0    mn6     ;30/2
XMN22 12 2  0    mn1     ; 4/2
C25 25 0 100f IC=1.8              ; Cin=10.3f
Vdd 98 0 DC 1.8
Vm1 98 99 DC 0
V1 1 0 PULSE(0     1.79   000p 200p 200p 300p 1000p)
V4 4 0 PULSE(1.80  0      140p 200p 200p 300p 1000p)
V2   2 0 PULSE(0    1.80 000p 200p 200p 500p 1000p)
.TRAN 1e-011 1e-009 0 1e-011
.TEMP 27
.PLOT TRAN V(1) V(2) V(4) V(25) 0,2
.end
```

Practice: Build and Test the CMOS NAND

Select V_{DD} = 5V, V_{SS} = 0V, 2N7000 NMOS transistors, and ZVP3306a PMOS transistors.

Transient Response of the circuit.

Connect the Function Generator square wave output to node y. Connect node z to a 10K resistor connected to 5V. Set the frequency to 10KHz. What do you see at node f? What did you expect to see?

Figure 920 NAND   **Figure 926 NAND Layout**

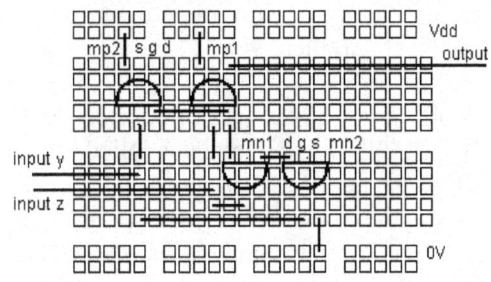

10 Semiconductor Diodes

10.1 Diode

The diode is a two-terminal device whose resistance depends upon the magnitude and polarity of the voltage across the diode terminals. In elementary terms, a modern forward-biased diode has a low resistance, while the resistance is very high when the diode is reversed biased.

Theory: The pn junction diode (Figure 1001) is represented by an equation for the diode current as a function of the applied voltage and diode physical parameters.

Figure 1001   q is charge of the electron (1.602×10^{-19} coulombs)
Diode in a circuit V_{pn} is voltage applied to the diode (volts)

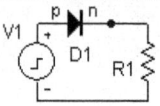

k is Boltzmann's constant (1.3805×10^{-23}
joules/degree Kelvin)
T is temperature (degrees Kelvin)

(1) $i_d = i_s(e^x - 1)$ $x = V_{pn}/V_T$

$V_T = kT/q = 25.85\,\text{mv}$ $T = 300°K$

When the normalized voltage x is more negative than −5 ($V_{PN} = -130\text{mv}$) the diode current is constant with value i_S (e.g. $i_S=10^{-15}A=1fA$). The diode is off for all practical purposes. At $V_{pn} = -130\text{mv}$ and with current $i_S=1fA=10^{-15}$ amperes, the resistance is about 10^{14} ohms. The normalizing voltage value V_T is approximately 26 millivolts at room temperature. When the normalized voltage x is positive, the diode is still off for all practical purposes in most circuits until $x=600\text{mV}/26\text{mV}=23$ ($e^{23}=9.75 \times 10^9$) and the diode conducts about $9.75E9 \times i_S$ or about 10 microamperes (Figure 100111).

Spice program 10011
```
Fig10011.ckt diode
V1 1 0 DC 0
D2 1 2 1N4148 OFF
R1 2 0 0
.Model 1N4148 D(IS=1f RS=16 CJO=2p TT=12n BV=100 IBV=400p)
.DC LIN V1 0 1 0.01
.PLOT DC I(V1) 0,-0.5M
.END
```

Figure 100111 Diode voltage-current vi constraint

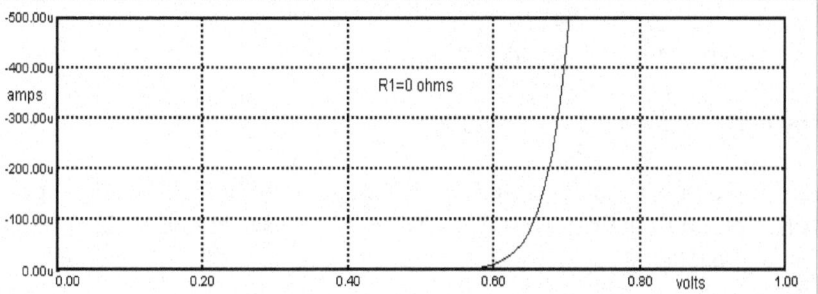

Practice *Rule of thumb* Analysis of most circuits is easier if you keep in mind that the pn junction diode is an open circuit for all practical purposes when the forward voltage falls below 0.6 volts, and the diode is like a closed switch when forward voltage is above 0.7 volts (Figure 100111). Forward means p is more positive than n (Figure 1002).

Figure 1002

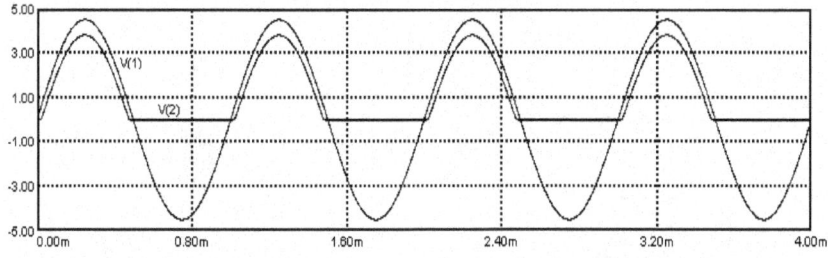

For an example of a diode application consider the half wave rectifier. The negative halves of the sine wave are deleted at V_2, because the diode is an open circuit (Figure 1002). The peak value of output V_2 is about 0.7 volts less than the input sine wave V_1.

Figure 100211 Half Wave Rectifier (Figure 1002)

```
Fig10021.ckt diode
V1 1 0 sin(0 4.5 1000 0 0}
D1 1 2 1N4148
R1 2 0 10000
.Model 1N4148 D(IS=1f RS=16 CJ0=2p TT=12n BV=100 IBV=400p)
.TRAN 1e-008 0.004 0 1e-006
.TEMP 27
.PLOT TRAN V(1) V(2)  -5,5
.END
```

117

Electronic Circuits – Practical Learning

Practice: Build and Test the Half Wave Rectifier

Select 1N4148 diode or equal. Select R1 = 10K.

Transient Response of the circuit.

Connect the Function Generator sine wave output to node V1. Select R1 = 10K. Set the frequency to 10KHz. What do you see at node V2? What did you expect to see?

Figure 1002 Half Wave Rectifier   **Figure 1003 Layout**

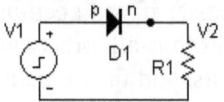

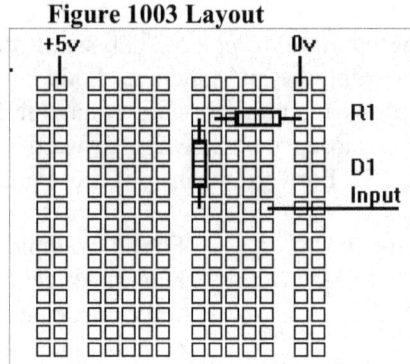

10.2 Zener Diode

Zener diode is a diode whose reverse breakdown voltage is designed to be V_Z volts, where one product line has diodes whose Zener voltage ranges from 3.3v to 100v.

Theory: In the breakdown region the current is usually limited by a series resistor.

Figure 1004   **Figure 1005 Zener Diode vi Constraint & loadline**

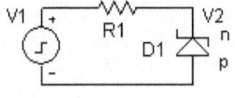

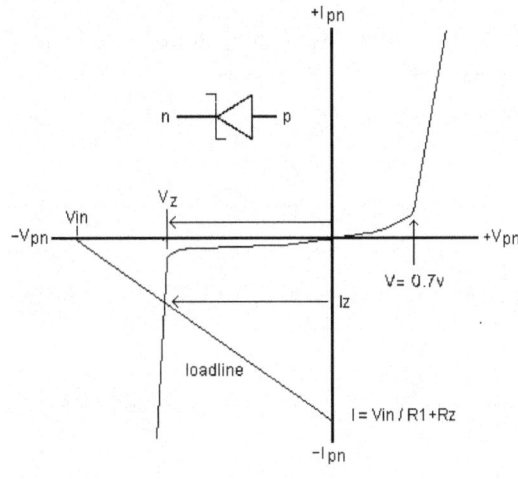

```
Fig10041.ckt zener diode
V1 1 0 pwl(0 0, .001 9, .001 0, 0.002 9}
R1 1 2 1000
D1 0 2 D1N4733

.model D1N4733 D(Is=1.214f Rs=1.078 Ikf=0 N=1 Xti=3 Eg=1.11
+ Cjo=185p M=.3509 Vj=.75 Fc=.5 Isr=2.601n Nr=2 Bv=5.1
Ibv=.70507 Nbv=.74348 Ibvl=4.8274m Nbvl=6.7393 Tbvl=176.471u)

*.PLOT TRAN V(1) V(2) 0,10
.TRAN 1e-008 0.002 0 1e-006
.TEMP 27
.PLOT TRAN I(V1) 0,-0.005
.END
```

Electronic Circuits – Practical Learning

A sawtooth waveform provides a convenient display of voltage and current as determined by the Zener diode.

As input V_1 increases no current flows until Zener voltage V_Z is reached, because the diode is reversed biased. As V_1 increases past V_Z current starts to flow (Figure 100412), because the Zener breaks down. As current starts to flow output voltage V_2 equals V_Z (Figure 100411).

The current $I = (V_1 - V_Z)/R_1$, where $V_1 - V_Z$ is the voltage across R_1. For example if $V_Z = 4$ volts, $V_1 = 9$ volts (Figure 100411), $R_1 = 1000$ ohms, then I $= (9-4)/1000 = 5$ma (Figure 100412).

Figure 100411 Output voltage limited by Zener

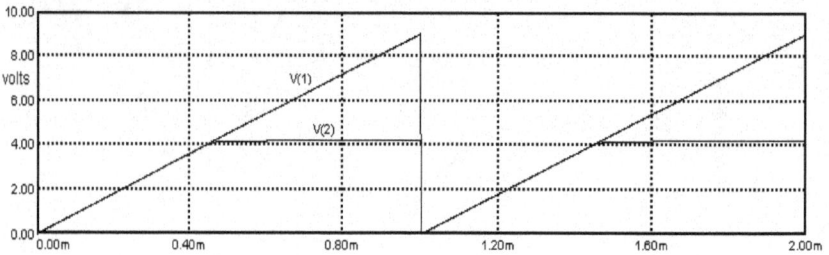

Figure 100412 Current I due to $(V_1 - V_Z)/R_1$

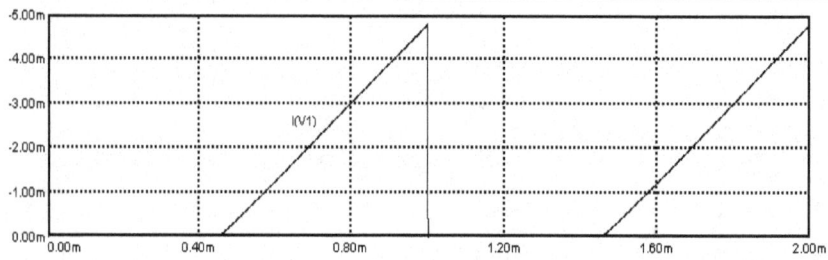

Practice: Build and Test the Half Wave Rectifier

Select D1N4733 diode or equal. Select R1 = 10K.

Transient Response of the circuit.

Connect the Function Generator sine wave output to node V1. Select R1 = 10K. Set the frequency to 10KHz. What do you see at node V2? What did you expect to see?

Figure 1004 Zener Circuit

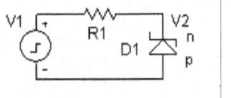

Figure 1006 Layout

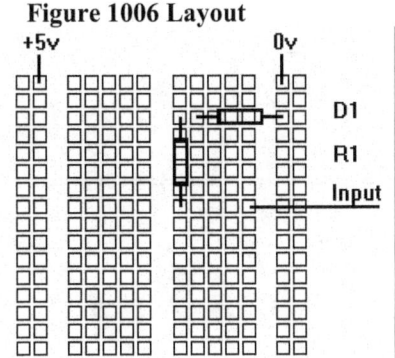

10.3 LED

A Light Emitting Diode (LED) produces visible light when current flows through it. The LED light is available in several colors such as red, green, and blue.

An LED emits light when turned on with a forward voltage drop of about 2 volts. A resistor in series limits the LED current to a safe value.

Figure 1007 LED

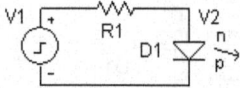

```
Fig10071.ckt led diode
V1 1 0 pwl(0 0, .001 5, .001 5, 0.002 0}
R1 1 2 330
D1 2 0 led1

.MODEL LED1 D (IS=93.2P RS=42M N=3.73 BV=4 IBV=10U
+ CJO=2.97P VJ=.75 M=.333 TT=4.32U)

*.PLOT TRAN I(V1) 0,-0.01
.TRAN 1e-008 0.002 0 1e-006
.TEMP 27
.PLOT TRAN V(1) V(2) 0,5
.END
```

A sawtooth waveform provides a convenient display of voltage and current as determined by the LED diode.

As input V_1 increases current flows when LED voltage $V_2 = 2$ volts is reached, because the diode is forward biased. As V_1 increases past 1.8V (Figure 100711) current increases (Figure 100712). As current starts to flow output voltage V_2 is limited to about 1.8v.

The current $I = (V_1–1.8)/R_1$, where $V_1–1.8$ is the voltage across R_1. For example if $V_1 = 5$ volts (Figure 1008), $R_1 = 330$ ohms, then $I = (5–1.8)/330 = 9.7$ma.

Figure 100711 LED Voltage V$_2$

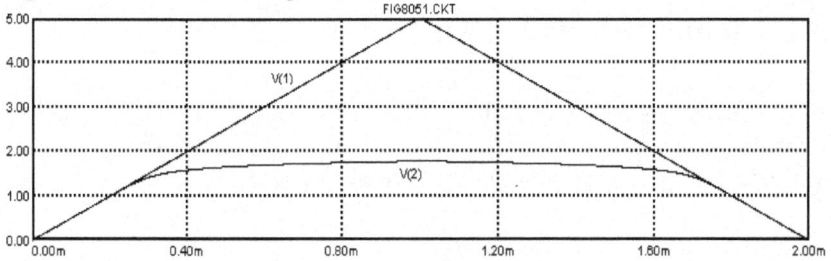

Figure 100712 LED Current I(V$_1$)

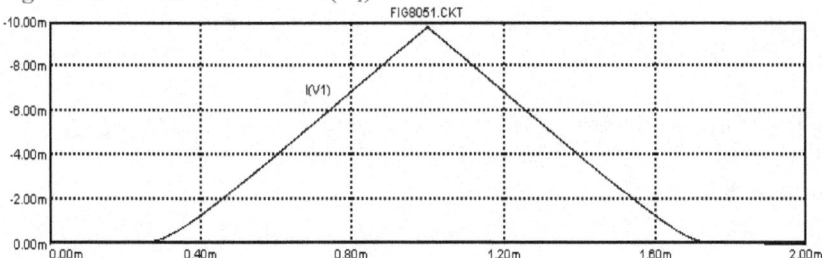

Figure 1008 LED Driver

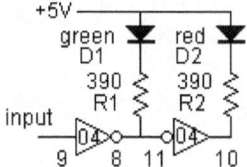

Figure 1009 LED Circuit

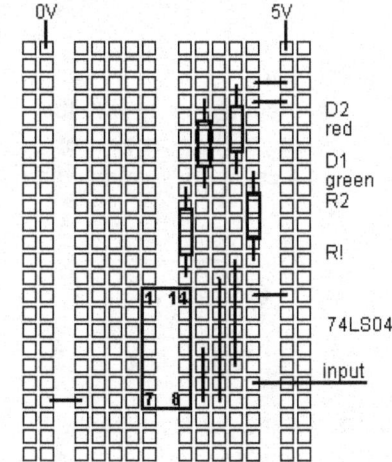

11 Circuit Analysis – Steady State

Electric circuit problems are solved by applying the *Kirchhoff connection constraint laws*, the *vi constraint laws*, the mesh or node methods that formulate equations, and Cramer's rule that solves the equations.

1) *Kirchhoff connection constraints*
Connection constraints impose conditions on voltages or currents in a circuit that depend on how the components are wired into the circuit. Connection constraints are independent of the specific types of components used.

KCL *The algebraic sum of all currents entering and leaving one node is zero.* The node method is based on Kirchhoff's Current Law (KCL)

Figure 1101 KCL

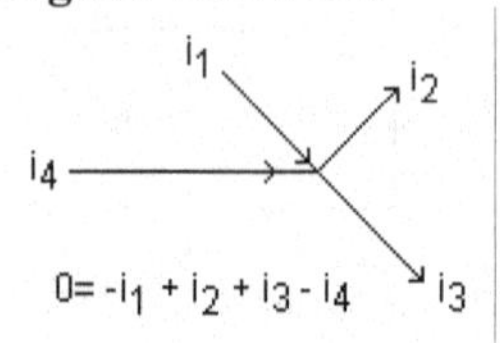

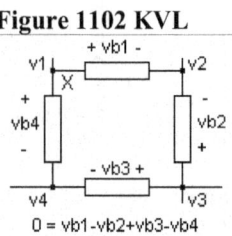

$0 = -i_1 + i_2 + i_3 - i_4$

KVL *The algebraic sum of the voltage differences around every closed circuit contour is zero.* The mesh method is based on Kirchhoff's Voltage Law (KVL)

Figure 1102 KVL

$0 = vb1 - vb2 + vb3 - vb4$

2) *Voltage-current constraints*
Ohm's law is the resistor vi constraint. The inductor vi constraint follows from a changing current flowing in a wire creating a changing flux that passes by its own (source) wire inducing a voltage in that same wire. From $Q = CV$ the differential equation relating a capacitor's voltage and current follows immediately.

(1) $\boxed{v(t) = Ri(t)}$ (2) $\boxed{v(t) = L\dfrac{di(t)}{dt}}$ (3) $\boxed{i(t) = C\dfrac{dv(t)}{dt}}$

3) *Formulate and solve Circuit Equations*
Formulate circuit mesh equations by using KVL and the mesh method. Formulate circuit node equations by using KCL and the node method. Solve by Cramer's rule

Cramer's Rule

The first subscript is the row number. The second subscript is the column number. Determinants are expanded by rows or columns.

Cramer's solutions are expansions by columns where forcing functions replace the column's elements. Note: incorporate minus signs into the a_{ij}'s.

Cramer found responses y_1, y_2 to forcing functions x_1, x_2.

$If\ x_1 = a_{11}y_1 + a_{12}y_2 \quad and \quad x_2 = a_{21}y_1 + a_{22}y_2$

$Then\ \Delta = a_{11}a_{22} - a_{21}a_{12} \quad and$

$$y_1 = \frac{\begin{vmatrix} x_1 & a_{12} \\ x_2 & a_{22} \\ a_{11} & a_{12} \\ a_{21} & a_{22} \end{vmatrix}}{} = \frac{x_1 a_{22} - x_2 a_{12}}{\Delta} \qquad y_2 = \frac{\begin{vmatrix} a_{11} & x_1 \\ a_{21} & x_2 \\ a_{11} & a_{12} \\ a_{21} & a_{22} \end{vmatrix}}{} = \frac{-x_1 a_{21} + x_2 a_{11}}{\Delta}$$

And, for three responses y_1, y_2, y_3 to forcing functions x_1, x_2, x_3.

$If\ x_1 = a_{11}y_1 + a_{12}y_2 + a_{13}y_3$

$\qquad x_2 = a_{21}y_1 + a_{22}y_2 + a_{23}y_3$

$\qquad x_3 = a_{31}y_1 + a_{32}y_2 + a_{33}y_3$

$Then\ \Delta = a_{11}\Delta_{11} - a_{21}\Delta_{21} + a_{31}\Delta_{31} \quad (expansion\ by\ column\ 1)$

$\Delta = a_{11}(a_{22}a_{33} - a_{23}a_{32}) - a_{21}(a_{12}a_{33} - a_{13}a_{32}) + a_{31}(a_{12}a_{23} - a_{13}a_{22})$

$$y_1 = \frac{x_1\Delta_{11} - x_2\Delta_{21} + x_3\Delta_{31}}{\Delta} \quad (expansion\ down\ column\ 1,\ rows\ 1,2,3)$$

$$y_2 = \frac{x_1\Delta_{12} - x_2\Delta_{22} + x_3\Delta_{32}}{\Delta} \quad (expansion\ down\ column\ 2,\ rows\ 1,2,3)$$

$$y_3 = \frac{x_1\Delta_{13} - x_2\Delta_{23} + x_3\Delta_{33}}{\Delta} \quad (expansion\ down\ column\ 3,\ rows\ 1,2,3)$$

11.1 Node Method

The node method *formulates* circuit equations in the form of sums of currents by applying Kirchhoff's current law (KCL) that states that the algebraic sum of all (branch) currents entering and leaving a node is zero. Here is our convention for *branch* currents.

(1a) $v_2 - v_1 = iz$

Figure 1103

(1b) $i = y(v_2 - v_1) \implies y = \dfrac{1}{z}$

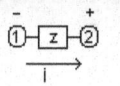

Our convention for KCL is that currents entering the node have a positive sign, and currents leaving the node have a negative sign.

Circuit Node Equations

There are five nodes that might produce equations (Figure 1104). However ground voltage $v_0 = 0$ volts, $v_{40} = v_4 - v_0 = v_s - 0$ so that $v_4 = v_s$. Consequently we need to find values for v_1, v_2, v_3. Observe that only current i_1 enters and leaves node 1. Since z_1 and z_2 are in series consider them to be one impedance z_{12}. This action eliminates node 1 for the moment. In other words we defer finding v_1 that equals $v_2 - i_1 z_2$. In this way we have reduced the problem to analysis of two nodes: 2 and 3.

Node 2 *Our convention is that we assume v_2 is greater than any other voltage when we formulate the node 2 equation.* Thus i_1, i_2, and i_3 flow into node 2 (Figure 1104). Their sum equals zero according to KCL. The sum (equation 2) is the KCL connection constraint.

Figure 1104 Node 2 Analysis

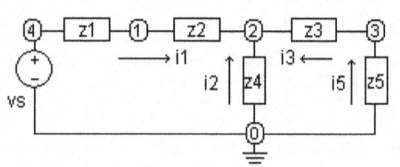

(2) $0 = i_1 + i_2 + i_3$

Each current equals a $y(v_a - v_b)$ expression that is found by inspection of Figure 1104, while applying equation 1b. So we start with sums of currents and end up with equations for node voltages.

(3a) $i_1 = y_{12}(v_2 - v_4)$ $i_2 = y_4(v_2 - v_0)$ $i_3 = y_3(v_2 - v_3)$

(3b) $0 = i_1 + i_2 + i_3 = y_{12}(v_2 - v_4) + y_4(v_2 - v_0) + y_3(v_2 - v_3)$

$0 = (y_{12} + y_4 + y_3)v_2 + y_{12}(-v_4) + y_4(-v_0) + y_3(-v_3)$

$0 = (y_{12} + y_4 + y_3)v_2 - (y_{12}v_4 + y_4v_0 + y_3v_3)$

Straightforward algebraic manipulations put the equation in *standard form* (equation 4b) with two unknowns v_2 and v_3. Observe that each admittance y is connected to node 2. That is why their sum is the coefficient of v_2.

(4a) $v_S = v_4$ and $v_0 = 0$

(4b) $y_{12}v_S = (y_{12} + y_4 + y_3)v_2 - y_3v_3$

Node 3 *Our convention is that we assume v_3 is greater than any other voltage when we formulate the node 3 equation 5.* Thus i_5 and i_3 flow into node 3 (Figure 1105).

Figure 1105 Node 3 Analysis

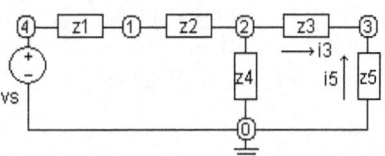

(5) $0 = i_3 + i_5$

Each current equals a $y(v_a - v_b)$ expression that is found by inspection of Figure 1105, while applying equation 1b. Again, note that *we assume v_3 is greater than any other voltage.*

(6a) $i_3 = y_3(v_3 - v_2)$ $i_5 = y_5(v_3 - v_0)$

(6b) $0 = y_3(v_3 - v_2) + y_5(v_3 - v_0)$

Algebraic manipulations put the equation in *standard form* (equation 7b) with two unknowns v_2 and v_3. Observe that each admittance y is connected to node 3 that is why their sum is the coefficient of v_3.

(7a) $v_0 = 0$

(7b) $0 = (y_3 + y_5)v_3 - y_3v_2$

Mathematics The two node equations (4b and 7b) with two unknowns v_2, v_3 are solved by using Cramer's rule.

(4b) $y_{12}v_S = (y_{12} + y_4 + y_3)v_2$ $- y_3v_3$

(7b) $0 =$ $- y_3v_2 + (y_3 + y_5)v_3$

There may be an independent or dependent current source in a branch. This is just another current at the two nodes connected to the current source. Add or subtract the current to the KCL node equations according to the current direction. However a voltage source in a branch is another matter (Section 11.2).

11.2 Node Method - Branch with Voltage Source

The voltage source in a branch can be an independent or dependent source. Assume the circuit model has an independent voltage source in one branch (Figure 1106). Formulating the sum of the currents at nodes 4 and 1 is straightforward (equations 8a and 8b). This is not so for nodes 2 and 3. What is the current through the voltage source v_s? One procedure for circuits like this is to *assume a value, such as* I_x, for the current flowing through the voltage source v_s. Then we can formulate the KCL equations for nodes 2 and 3 (equations 8c and 8d), as well as a fifth equation 8e relating v_2 to v_3. The I_x trick is not obvious, however it solves the problem. Note: Figure 1106 circuit is NOT related to Figure 1105 circuit!

Figure 1106 Branch with voltage source

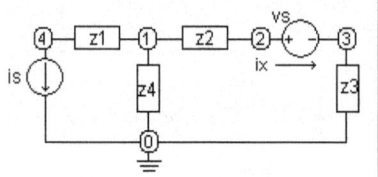

(8a) node 4 $i_s = y_1(v_4 - v_1)$

(8b) node 1 $0 = (y_1 + y_2 + y_4)v_1 - y_1v_4 - y_2v_2$

(8c) node 2 $0 = -y_2v_1 + y_2v_2 - I_x$

(8d) node 3 $0 = y_3v_3 + I_x$

(8e) $v_2 = v_s + v_3$

Observe that I_x drops out when the v_2 and v_3 node equations are added. Addition leaves three equations. We still need a fourth equation. The fourth equation is equation 8e that relates the voltage source v_s to node voltages v_2 and v_3.

(8a)   node 4 $i_s = y_1(v_4 - v_1)$

(8b)   node 1 $0 = (y_1 + y_2 + y_4)v_1 - y_1v_4 - y_2v_2$

(8c + 8d)   node 2+3 $0 = -y_2v_1 + y_2v_2 + y_3v_3$

(8e) $v_2 = v_s + v_3$

Substitute 8e into 8c+8d to eliminate v_2. Now we have three equations and three unknowns. The rest is mathematics.

(8a)   node 4 $i_s = y_1v_4 - y_1v_1$

(8b)   node 1 $y_2v_s = (y_1 + y_2 + y_4)v_1 - y_1v_4 - y_2v_3$

(8c + 8d,8e) node2+3 $-y_2v_s = -y_2v_1 + (y_2 + y_3)v_3$

128

11.3 Maxwell's Mesh Current Method

The mesh method formulates circuit equations in the form of sums of voltages by applying Kirchhoff's voltage law (KVL) that states that the algebraic sum of all (branch) voltages around a mesh is zero. Here is our convention for *branch* voltages.

(1a) $v_2 - v_1 = iz$

(1b) $i = y(v_2 - v_1) \implies y = \dfrac{1}{z}$

Figure 1103

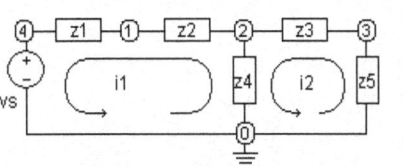

Circuit mesh equations

Mesh 1 Circling mesh 1 we pass nodes 4, 1, 2, 0, and return to node 4. The voltage differences are v_{4-1}, v_{1-2}, v_{2-0}, and v_{0-4} whose sum equals zero according to KVL (equation 9). The sum is the KVL connection constraint.

Figure 1107 Mesh analysis

For convenience let $v_{pq} = v_p - v_q$ where p and q are node numbers.

(9) $0 = v_{41} + v_{12} + v_{20} + v_{04}$

Each voltage difference v_{pq} equals an iz expression that is found by inspection of Figure 1107, while applying equation 1a.

(10a) $v_{41} = z_1 i_1 \quad v_{12} = z_2 i_1 \quad v_{20} = z_4(i_1 - i_2) \quad v_{04} = -v_s$

(10b) $0 = v_{41} + v_{12} + v_{20} + v_{04}$

(10c) $0 = z_1 i_1 + z_2 i_1 + z_4(i_1 - i_2) - v_s$

(10d) $0 = (z_1 + z_2 + z_4)i_1 - (z_4)i_2 - v_s$

Straightforward algebraic manipulations put the equation in *standard form* (equation 10) with two unknowns i_1 and i_2. Observe that each impedance z is in mesh 1 that is why their sum is the coefficient of i_1. So we start with sums of voltages and end up with equations for mesh currents.

(11) $v_s = (z_1 + z_2 + z_4)i_1 - z_4 i_2$

129

Mesh 2 Circling mesh 2 we pass nodes 2, 3, 0, and return to node 2. The voltage differences are v_{23}, v_{30}, and v_{02} whose sum equals zero according to KVL (equation 12). The sum is the KVL connection constraint.

Figure 1107 Mesh analysis

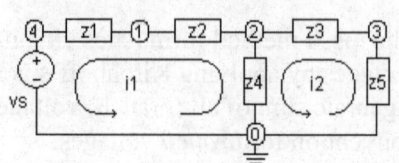

(12) $0 = v_{23} + v_{30} + v_{02}$

Each voltage difference v_{pq} equals an iz expression that is found by inspection of Figure 1107.

(13a) $v_{23} = z_3 i_2$ $v_{30} = z_5 i_2$ $v_{02} = z_4(i_2 - i_1)$

(13b) $0 = z_3 i_2 + z_5 i_2 + z_4(i_2 - i_1)$

Straightforward algebraic manipulations put the equation in *standard form* (equation 14) with two unknowns i_1 and i_2. Observe that each impedance z in mesh 2 is part of the sum that is the coefficient of i_2.

(14) $0 = -z_4 i_1 + (z_3 + z_4 + z_5) i_2$

Mathematics The two node equations (11 and 14) with two unknowns i_1, i_2 are solved by using Cramer's rule. With experience you will be able to write these down immediately.

(11) $v_s = (z_1 + z_2 + z_4) i_1 - z_4 i_2$

(14) $0 = -z_4 i_1 + (z_3 + z_4 + z_5) i_2$

There can be an independent or dependent voltage source in a branch. This is just another voltage in the mesh. Add or subtract the voltage to the KVL mesh equations according to the voltage polarity. However a current source in a branch is another matter (Section 11.4).

11.4 Mesh Method – Branch with a Current Source

A branch may include an independent or dependent current source. A circuit model for a common emitter transistor circuit (Figure 1108) has a dependent current source in one branch. Formulating the sum of the voltages around mesh 1 is straightforward. This is not so for meshes 2 and 3. What is the voltage drop v_x across the dependent current source? It is not an

Figure 1108 Branch with a dependent current source

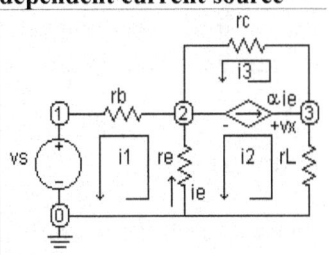

iR drop. The *trick* we use is to assume a value, such as v_x, for the voltage difference across the current source so that we can proceed directly to formulate the mesh equations. Equation 15d relates the dependent current source to the mesh currents

(15a) $v_s = (r_e + r_b)i_1 - r_e i_2$

(15b) $0 = -r_e i_1 + (r_e + r_L)i_2 - v_x$

(15c) $0 = r_c i_3 + v_x$

(15d) $i_e = i_1 - i_2$

Observe that v_x drops out when the i_2 and i_3 mesh equations 15b and 15c are added. Addition leaves two equations. We still need a third equation. The third equation is equation 15d that relates the dependent current source i_e to mesh currents i_1 and i_2.

(15a) $v_s = (r_e + r_b)i_1 - r_e i_2$

(15b+15c) $0 = -r_e i_1 + (r_e + r_L)i_2 + r_c i_3$

(15d) $i_e = i_1 - i_2$

We can eliminate i_3. The KCL equation at node 3 is equation 16a.

(16a) $\alpha i_e = i_3 - i_2$

(16b) $i_3 = i_2 + \alpha i_e = i_2 + \alpha(i_1 - i_2)$

(16c) $i_3 = \alpha i_1 + (1 - \alpha)i_2$

Substitute (16c) into (15b+15c). Now we have two equations and two unknowns. The rest is mathematics.

(15a) $v_s = (r_e + r_b)i_1 - r_e i_2$

(15b+15c,16c) $0 = (-r_e + r_c \alpha)i_1 + [r_e + r_L + r_c(1 - \alpha)]i_2$

Electronic Circuits – Practical Learning

11.5 Example: Low Pass Filter

An ideal low pass filter passes sine waves whose frequency is less than the "cutoff" frequency with zero attenuation, and attenuates to zero sine waves whose frequency is greater than the "cutoff" frequency. A real filter approximates this performance.

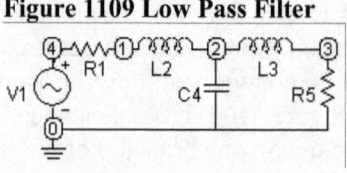

mesh method The vi constraints of filter impedances (Figure 1109) when substituted in equations 11 and 14 produce equations 20.

(11) mesh 1 $v_s = (z_1 + z_2 + z_4)i_1 \qquad - z_4 i_2$

(14) mesh 2 $0 = \qquad\quad - z_4 i_1 + (z_3 + z_4 + z_5)i_2$

(20a) mesh 1 $v_s = \left(R_1 + pL_2 + \dfrac{1}{pC_4}\right)i_1 - \dfrac{1}{pC_4}i_2$

(20b) mesh 2 $0 = -\dfrac{1}{pC_4}i_1 + \left(pL_3 + \dfrac{1}{pC_4} + R_5\right)i_2$

node method The vi constraints of filter admittances (Figure 1109) when substituted in equations 4b and 7b produce equations 21.

(4b) node 2 $y_{12}v_s = (y_{12} + y_4 + y_3)v_2 \qquad - y_3 v_3$

(7b) node 3 $0 = \qquad\qquad - y_3 v_2 + (y_3 + y_5)v_3$

(21a) node 2 $\dfrac{1}{R_1 + pL_2}v_s = \left(\dfrac{1}{R_1 + pL_2} + pC_4 + \dfrac{1}{pL_3}\right)v_2 - \dfrac{1}{pL_3}v_3$

(21b) node 3 $0 = -\dfrac{1}{pL_3}v_2 + \left(\dfrac{1}{pL_3} + \dfrac{1}{R_5}\right)v_3$

Solution of Node Equations. Low Pass Filter

Use Cramer's rule to solve the node equations. The two node general solution proceeds as follows.

(22) node 2 $i_S = y_{22}v_2 - y_{23}v_3$

node 3 $0 = -y_{32}v_2 + y_{33}v_3$

(23) $\Delta_Y = y_{22}y_{33} - y_{23}y_{32}$

$$(24a) \quad v_2 = \frac{\begin{vmatrix} i_S & -y_{23} \\ 0 & y_{33} \end{vmatrix}}{\Delta_Y} \qquad (24b) \quad v_3 = \frac{\begin{vmatrix} y_{22} & i_S \\ -y_{32} & 0 \end{vmatrix}}{\Delta_Y}$$

Low Pass Filter Solution

Figure 1109 Low Pass Filter

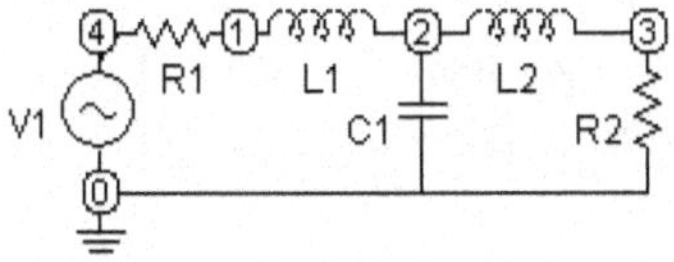

let $R = R_1 = R_2$, $L = L_1 = L_2$, $C = C_1$

$$(21a) \quad \text{node 2} \quad i_1 = \frac{1}{R+pL}v_1 = \left(\frac{1}{R+pL} + pC + \frac{1}{pL}\right)v_2 - \frac{1}{pL}v_3$$

$$(21b) \quad \text{node 3} \qquad\qquad 0 = -\frac{1}{pL}v_2 + \left(\frac{1}{pL} + \frac{1}{R}\right)v_3$$

$$(25) \quad \Delta_Y = \left(\frac{1}{R+pL} + pC + \frac{1}{pL}\right)\left(\frac{1}{pL} + \frac{1}{R}\right) - \left(\frac{1}{pL}\right)^2 = \frac{1}{pLR}\left(2 + pCR + p^2LC\right)$$

$$(26) \quad v_2 = \frac{\begin{vmatrix} i_S & -y_{23} \\ 0 & y_{33} \end{vmatrix}}{\Delta_Y} = \frac{\dfrac{1}{R+pL}v_S \cdot \left(\dfrac{1}{pL} + \dfrac{1}{R}\right)}{\dfrac{1}{pLR}\left(2 + pCR + p^2LC\right)} = \frac{v_S}{\left(2 + pCR + p^2LC\right)}$$

$$(27) \quad v_3 = \frac{\begin{vmatrix} y_{22} & i_S \\ -y_{32} & 0 \end{vmatrix}}{\Delta_Y} = \frac{\dfrac{1}{R+pL}v_S \cdot \dfrac{1}{pL}}{\dfrac{1}{pLR}\left(2 + pCR + p^2LC\right)} = \frac{Rv_S}{\left(R+pL\right)\left(2 + pCR + p^2LC\right)}$$

133

Electronic Circuits – Practical Learning

if $z_C = z_L$ at ω_0, then $\omega_0 L = \dfrac{1}{\omega_0 C}$ or $LC = \dfrac{1}{\omega_0^2}$

$$CR = \frac{\omega_0}{\omega_0} CR = \omega_0 CR \frac{1}{\omega_0} = \frac{R}{\omega_0 L} \frac{1}{\omega_0} = \frac{1}{Q_S} \frac{1}{\omega_0} \text{ and } \frac{L}{R} = \frac{\omega_0 L}{R} \frac{1}{\omega_0} = Q_S \frac{1}{\omega_0}$$

(28) $\quad T(p) = \dfrac{1}{\left(\dfrac{p^2}{\omega_0^2} + \dfrac{1}{Q_S}\dfrac{p}{\omega_0} + 2\right)\left(1 + Q_S\dfrac{p}{\omega_0}\right)}$

Select values for a specific design.
If $L = 470\mu H$ and $C = 1000pF$, then

(29) $\quad Z_0 = \omega_0 L = \dfrac{L}{\sqrt{LC}} = \sqrt{\dfrac{L}{C}} = \sqrt{\dfrac{470u}{1000p}} = 685\Omega$

(30) $\quad f_0 = \dfrac{\omega_0}{2\pi} = \dfrac{1}{2\pi}\dfrac{1}{\sqrt{LC}} = \dfrac{1}{2\pi}\dfrac{1}{\sqrt{470\times10^{-6}\times1000\times10^{-12}}} = 232KHz$

(31) $\quad Q = \dfrac{Z_0}{R} \rightarrow Q = \dfrac{685}{330} = 2.1, \quad Q = \dfrac{685}{680} = 1, \quad Q = \dfrac{685}{1500} = 0.46$

Spice program 11091 calculates the frequency response (equation 28).
Voltage v drives three low pass filters in parallel to show how resistor values affect the transfer function.

```
Fig11091.ckt   low pass filter
V 1 0 AC 1 0 PULSE(0 4 0 0 0 1000u 2000u)
*V 1 0 AC 1 0 PULSE(0 4 0 0 0 10u 20u)
R1 1 2 1500        ;Q=0.46
L1 2 3 470u
L2 3 4 470u
C1 3 0 1000p
R2 4 0 1500
R11   1 12 680     ;Q=1
L11 12 13 470u
L12 13 14 470u
C11 13  0 1000p
R12 14  0 680
R21   1 22 330     ;Q=2.1
L21 22 23 470u
L22 23 24 470u
C21 23  0 1000p
R22 24  0 330
.AC DEC 200 1000 1e+007
.PLOT AC VDB(4) VDB(14) VDB(24) -50,0
.TRAN 2e-008 2e-005 0 1e-008
.TEMP 27
.PLOT TRAN V(4) V(14) V(24) 0,2.5
.end
```

Figure 110911 Low Pass Filter Frequency Response

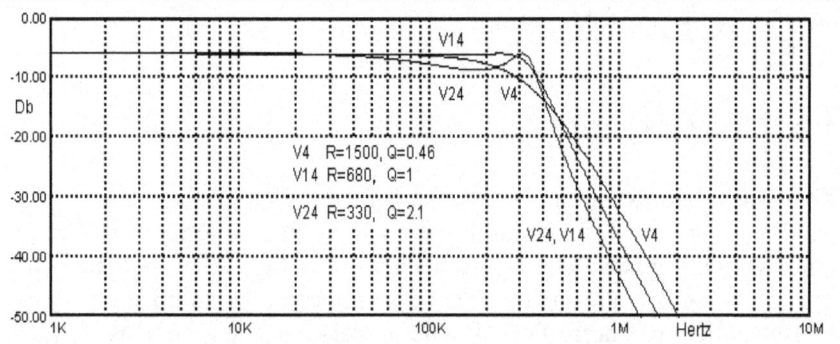

Spice program 11091 also calculates transient response when we add TRAN and PLOT TRAN lines. Note how the plots depend on R.

```
Fig11091.ckt  low pass filter
V 1 0 AC 1 0 PULSE(0 4 0 0 0 1000u 2000u) ; Figure 110712
*V 1 0 AC 1 0 PULSE(0 4 0 0 0 10u 20u)    ; Figure 110713
.TRAN 2e-008 2e-005 0 1e-008
.PLOT TRAN V(4) V(14) V(24) 0,2.5
```

Figure 110912 Low Pass Filter Transient Response to Step Function

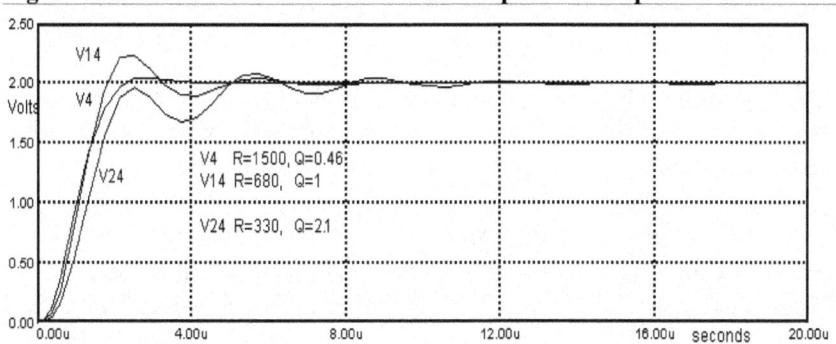

Figure 110913 Low Pass Filter Transient Response to 10µs Pulse

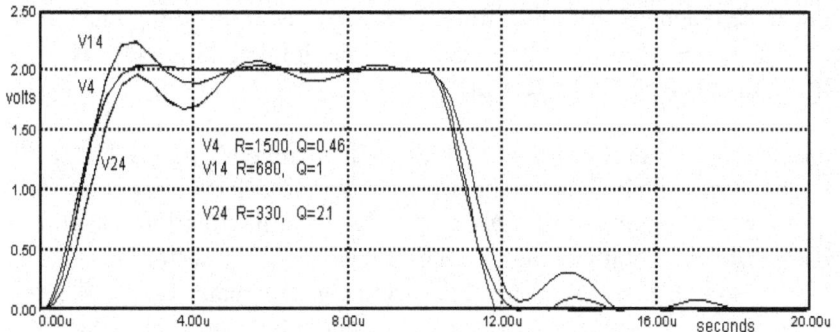

135

Appendix

A1 Complex Numbers

The words complex and imaginary are potentially misleading, because complex numbers are not complicated and imaginary operators are not part of someone's imagination. Both words are labels: they are technical terms used to designate a class of numbers. A complex number z is represented by an ordered pair of real numbers x and y written as (x, y).

Multiplication by −1 and √−1 A number can be represented as a distance on a number line. We define steps to the right as positive so that distance AB=+4. Multiply +4 by −1 to get −4 that is the distance AC. Multiply AC by −1 to get back to AB. Clearly multiplication by −1 in effect *rotates* AB and AC by 180°.

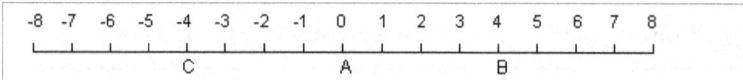

If +4 is multiplied by √−1 the result is 4√−1. Multiply 4√−1 by √−1 to get −4. Hence multiplication by √−1 two times rotates AB by 180°. And so multiplication by √−1 implements a 90° rotation of AB.

The world has agreed that numbers such as 4√−1 are *imaginary* numbers. To save writing √−1 is replaced by i in the mathematical literature. However EE's use i to designate current. That is why they use j for √−1.

Complex numbers The ordered pair (x_1, y_1) is a point in the (x, jy) plane that can be reached by starting from the origin, marching along the x-axis for a distance x_1, rotating $\pi/2$ radians, and marching parallel to the jy-axis for distance y_1 (Figure CN1a).

Working with ordered pairs (x, y) does not have much appeal, which is why the world adopted the well known alternative z=x+jy that is easier to work with. In other words: taking our clue from the rotation operation we use j as a $\pi/2$ rotation operator. Then we say jy_1 is a vector we add to vector x_1 so that $z_1=x_1+jy_1$. This replaces the ordered pair (x_1, y_1). We say z is a complex number whose real part is x and whose imaginary part is y. Keep in mind that x and y are real numbers.

Figure CN1 Complex numbers in Cartesian and polar coordinates

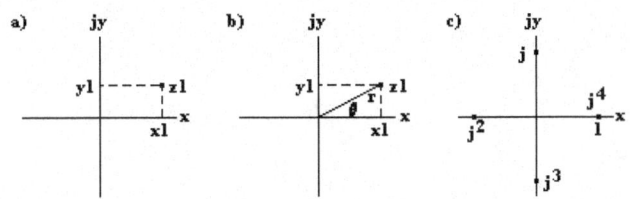

Polar coordinates: If r is the distance from the origin to the point z, then x = r cos θ, and y = r sin θ (Figure CN1b). See Euler relation below.

(1) $\quad z = x + jy = r\cos\theta + jr\sin\theta = re^{j\theta}$

(2) $\quad \tan\theta = \dfrac{y}{x}$ so that $\theta = \tan^{-1}\dfrac{y}{x}$

Multiples of j Representing j as a π/2 rotation yields the same results as the √-1 representation (Figure CN1c, Euler).

(3) $\quad j = e^{j\frac{\pi}{2}} = \cos\dfrac{\pi}{2} + j\sin\dfrac{\pi}{2} = 0 + j1 = j$

(4) $\quad j^2 = e^{j\frac{\pi}{2}2} = e^{j\pi} = \cos\pi + j\sin\pi = -1 + j0 = -1$

(5) $\quad j^3 = e^{j\frac{\pi}{2}3} = e^{j\frac{3\pi}{2}} = \cos\dfrac{3\pi}{2} + j\sin\dfrac{3\pi}{2} = -0 - j1 = -j$

(6) $\quad j^4 = e^{j\frac{\pi}{2}4} = e^{j2\pi} = \cos2\pi + j\sin2\pi = 1 + j0 = 1$

Addition The sum of complex numbers is found by adding the two x's and then the y's.

$$z_1 + z_2 = (x_1 + jy_1) + (x_2 + jy_2)$$

(7) $z_1 + z_2 = (x_1 + x_2) + j(y_1 + y_2)$

Multiplication Find the product $z_1 z_2$. To find it multiply z_1 and z_2, while treating j as another real number. Then substitute −1 for j^2.

(8) $\quad z_1 z_2 = (x_1 + jy_1)(x_2 + jy_2)$

$$= x_1 x_2 + x_1 jy_2 + jy_1 x_2 + jy_1 jy_2$$
$$= x_1 x_2 + jy_1 jy_2 + jy_2 x_1 + jy_1 x_2$$
$$= x_1 x_2 + j^2 y_1 y_2 + j(x_2 y_1 + x_1 y_2)$$
$$= (x_1 x_2 - y_1 y_2) + j(x_2 y_1 + x_1 y_2)$$

137

Subtraction Subtraction is defined as addition of positive and negative complex numbers.

$$(9) \quad z_1 - z_2 = z_1 + [-z_2] = (x_1 + y_1) + (-x_2 - jy_2)$$
$$= (x_1 - x_2) + j(y_1 - y_2)$$

Division Division is facilitated by the complex conjugate concept, where j is replaced by –j.

$$\textit{If } z = x + jy, \text{ then } \bar{z} = x - jy$$

$$z\bar{z} = (x + jy)(x - jy) = x^2 - j^2 y^2 + jxy - jyx$$

$$(10) \quad z\bar{z} = x^2 + y^2 = r^2 = |z|^2 = |z| \times |z|$$

$$\frac{z_1}{z_2} = \frac{z_1}{z_2} \times \frac{\bar{z}_2}{\bar{z}_2} = \frac{(x_1 + jy_1)(x_2 - jy_2)}{r_2^2} = \frac{x_1 x_2 - j^2 y_1 y_2 - jx_1 y_2 + jy_1 x_2}{r_2^2}$$

$$(11) \quad \frac{z_1}{z_2} = \frac{x_1 x_2 + y_1 y_2}{r_2^2} + j\frac{x_2 y_1 - x_1 y_2}{r_2^2}$$

Euler Relation (Figure CN1b)

$$\textit{If } r = 1 \quad \textit{then} \quad z = \cos\theta + j\sin\theta$$

$$\frac{dz}{d\theta} = -\sin\theta + j\cos\theta = j(\cos\theta + j\sin\theta) = jz$$

$$(12) \quad \frac{dz}{z} = jd\theta$$

Integrating $\quad \ln z = j\theta + constant$

If $\theta = 0$ *then* $z = 1$ *so that* $\ln 1 = j0 + constant$

However, $\ln 1 = 0$ *so that constant* $= 0$

$$\therefore \ \ln z = j\theta \ \Rightarrow \ z = e^{j\theta}$$

$$(13) \quad e^{j\theta} = \cos\theta + j\sin\theta$$

A2 Laplace Transform

Transient response of electric circuits requires solution of ordinary and partial differential equations

The Laplace Transform is a terrific tool for solving ordinary and partial differential equations, because it produces the frequency response equations, including initial conditions, as well as the transient response equations. The Laplace transform transforms the ordinary differential equations of electric circuits into elementary algebraic equations. The algebraic equations are manipulated to solve for the variables of interest. Then the manipulated equations are inverse transformed back to the time domain *as a solution* of the original problem represented by the differential equations.

The time domain equations of a resistor-only circuit are algebraic. Resistors do not store energy, and so there are no transient states that dissipate stored energy in a resistor-only circuit. There are no initial conditions, because there are no constants of integration.

However, when energy storing capacitors and, inductors are included in a circuit everything changes.

Circuit equation (1) of an RLC circuit (Figure 701) is an integro-differential equation. Terms in the time domain equation are no longer exclusively algebraic; some terms may include differentials and others may include integrals.

Figure 701 Series RLC circuit

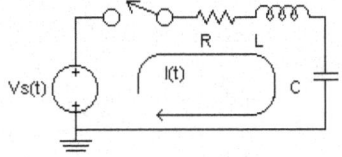

$$(1) \quad v_S(t) = Ri(t) + L\frac{di}{dt} + \frac{1}{C}\int_0^t i(x)dx$$

Two initial conditions at time t=0 are possible, because there are two constants of integration. Inductor L stores energy when a current flows through it. Capacitor C stores energy when a voltage is across it. The $v_S(t)$ waveform is not restricted. You are faced with the apparently formidable problem of finding solutions for simultaneous sets of such equations, which the Laplace transform provides an amazingly easy way to solve.

Electronic Circuits – Practical Learning

A2.1 The Laplace Transform

Let $f(t)$ be a real or complex valued function of time t for t>0, and $p=\sigma+j\omega$ be a complex variable used as a parameter. Then the Laplace transform of $f(t)$ is defined as

$$(2) \quad F(p)=\mathcal{L}[f(t)]=\int_0^\infty f(t)e^{-pt}\,dt$$

The symbol $\mathcal{L}$ is the Laplace transformation *operator* that invokes the Laplace transform. Equations in the *time domain variable t* are transformed into equations in the *complex frequency domain variable p*.

Inverse transform If we use the Laplace transform to solve a problem in the p domain, then we need an inverse transform to return to the time domain.

$$(3) \quad f(t)=\mathcal{L}^{-1}[F(p)]=\frac{1}{2\pi i}\int_{\sigma-j\infty}^{\sigma+j\infty}F(p)e^{tp}\,dp$$

A return from the *complex frequency domain p* to the time domain t is achieved by performing the inverse operation. The operation is known as the Inverse Laplace Transform (equation 3).

This integral for calculating the inverse transform is a "contour integration in the p plane." This process is part of the mathematical theory of functions of complex variables that is presented in mathematics texts.[1]

We use a well known trick that allows us to avoid calculating this formidable integral for almost all problems we might ever encounter.

The trick is straightforward. Expand the algebraic solution into a sum of partial fractions (Appendix A1). Then use the inverse of the transform of the exponential.

$$(4) \quad \mathcal{L}[e^{-at}]=\frac{1}{p+a} \quad \Rightarrow \quad e^{-at} \Leftrightarrow \frac{1}{p+a}$$

---

[1] Joel L. Shiff, *The Laplace Transform*, ISBN 0387 986 987

A2.2 Transforms Simplify Functions

Many complicated functions of real variable
t directly transform into elementary
functions of a complex variable p.

Figure 703 Step V_m u(t)

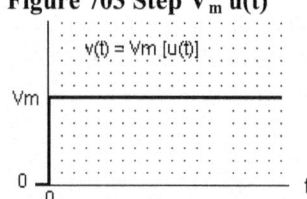

Transform of the step function u(t) The
unit step function (Figure 703,) transforms
to the algebraic expression $1/p$.

$$u(t) = 1 \quad for \ all \ t$$

$$F(p) = \mathcal{L}[u(t)] = \int_0^\infty u(t)e^{-pt}dt = \int_0^\infty 1 \times e^{-pt}dt = \frac{e^{-pt}}{-p}\Big|_0^\infty = \frac{0}{-p} - \frac{e^{-0}}{-p} = \frac{1}{p}$$

where $e^{-\infty} = 0$ means $\lim_{t\to\infty} e^{-pt} = 0$

$$(5) \quad u(t) \Leftrightarrow \frac{1}{p}$$

You can also read the u(t) transform pair as "one over p is the Laplace
transform of one", or "one is the Inverse Laplace transform of one over
p".

Transform of exp(–at) Most transcendental functions transform into
algebraic expressions. Consider the exponential function.

$$F(p) = \mathcal{L}[e^{-at}u(t)] = \int_0^\infty e^{-at}u(t)e^{-pt}dt = \int_0^\infty e^{-(p+a)t}dt = \frac{e^{-(p+a)t}}{-(p+a)}\Big|_0^\infty$$

$$= \frac{0}{-(p+a)} - \frac{e^{-0}}{-(p+a)} = \frac{1}{p+a}$$

$$(6) \quad e^{-at}u(t) \Leftrightarrow \frac{1}{p+a}$$

The transform of u(t) follows from this result by letting a = 0. This is a
useful method for generating other transform pairs.

Problem 1 From the definition of the Laplace transform calculate
a) $F(p) = \mathcal{L}[f(t)]$ for $f(t) = 3t$
b) $F(p) = \mathcal{L}[f(t)]$ for $f(t) = 7te^{-3t}$
c) $F(p) = \mathcal{L}[f(t)]$ for $f(t) = \cosh \omega t$
d) $F(p) = \mathcal{L}[f(t)]$ for $f(t) = t + e^{-at}$

141

Transforms of damped sin ωt and cos ωt The sin and cos functions in exponential format reveal that the transform of an exponential can be used as a shortcut.

Damped sine function

$$F(p) = \mathcal{L}[e^{-\sigma t} \sin \omega t \times u(t)]$$

$$F(p) = \int_0^\infty \frac{e^{-(\sigma-i\omega)t} - e^{-(\sigma+i\omega)t}}{2i} u(t) e^{-pt} dt = \int_0^\infty \frac{e^{-(p+\sigma-i\omega)t} - e^{-(p+\sigma+i\omega)t}}{2i} dt$$

$$F(p) = \frac{1}{2i}\left(\frac{1}{p+\sigma-i\omega} - \frac{1}{p+\sigma+i\omega}\right) = \frac{\omega}{(p+\sigma)^2 + \omega^2}$$

(7) $e^{-\sigma t} \sin \omega t \times u(t) \Leftrightarrow \dfrac{\omega}{(p+\sigma)^2 + \omega^2}$

Damped cosine function

$$F(p) = \mathcal{L}[e^{-\sigma t} \cos \omega t \times u(t)]$$

$$F(p) = \int_0^\infty \frac{e^{-(\sigma-i\omega)t} + e^{-(\sigma+i\omega)t}}{2} u(t) e^{-pt} dt = \int_0^\infty \frac{e^{-(p+\sigma-i\omega)t} + e^{-(p+\sigma+i\omega)t}}{2} dt$$

$$F(p) = \frac{1}{2}\left(\frac{1}{p+\sigma-i\omega} + \frac{1}{p+\sigma+i\omega}\right) = \frac{p+\sigma}{(p+\sigma)^2 + \omega^2}$$

(8) $e^{-\sigma t} \cos \omega t \times u(t) \Leftrightarrow \dfrac{p+\sigma}{(p+\sigma)^2 + \omega^2}$

Transform of the ramp function We have no shortcuts for the ramp function integration. Here we integrate by parts.

$$ramp: \ F(p) = \mathcal{L}[\frac{t}{T} u(t)] = \int_0^\infty \frac{t}{T} u(t) e^{-pt} dt = \frac{1}{T} \int_0^\infty t \times e^{-pt} dt = \frac{1}{T} \int_0^\infty u \, dv$$

$$let \quad u = t, \ dv = e^{-pt} dt. \quad \Rightarrow \quad du = dt, v = -\frac{1}{p} e^{-pt}$$

$$F(p) = \frac{1}{T} \int_0^\infty u \, dv = \frac{1}{T} uv \Big|_0^\infty - \frac{1}{T} \int_0^\infty v \, du$$

$$F(p) = -\frac{1}{T} t \frac{1}{p} e^{-pt} \Big|_0^\infty + \frac{1}{pT} \int_0^\infty e^{-pt} dt = 0 + \frac{1}{pT}\frac{-1}{p} e^{-pt} \Big|_0^\infty = \frac{1}{T}\frac{1}{p^2}$$

(9) $tu(t) \Leftrightarrow \dfrac{1}{p^2}$

A2.3 Transforms Simplify Operations

The vi constraints for R, L, and C components include multiplication by a constant, differentiation, and integration. We need the Laplace transforms of these operations. We use integration by parts to find the transform of a derivative and an integral. However, first we show that multiplication by a constant carries over from the real domain to the complex domain. This is relevant, because constants such as R, L, and C are coefficients of terms.

Transform of multiplication by a constant

if $f(t) \Leftrightarrow F(p)$, then

$$\mathcal{L}[K \cdot f(t)] = \int_{o}^{\infty} K \cdot f(t)e^{-pt}dt = K \cdot \int_{0}^{\infty} f(t)e^{-pt}dt = K \cdot F(p)$$

(10) $Kf(t) \Leftrightarrow KF(p)$

Transform of a first derivative

if $f(t) \Leftrightarrow F(p)$ and $\dfrac{df(t)}{dt}u(t) \Leftrightarrow F_1(p)$, then

$$F_1(p) = \mathcal{L}\left[\frac{df(t)}{dt}u(t)\right] = \int_{0}^{\infty}\frac{df(t)}{dt}u(t)e^{-pt}dt = \int_{0}^{\infty}\frac{df(t)}{dt}e^{-pt}dt$$

if $u = e^{-pt}, dv = \dfrac{df(t)}{dt}dt$ then $du = -pe^{-pt}dt$ $v = f(t)$

integrating by parts: $F_1(p) = \int_{0}^{\infty}udv = uv\Big|_{0}^{\infty} - \int_{0}^{\infty}vdu$

$$F_1(p) = e^{-pt}f(t)\Big|_{0}^{\infty} - \int_{0}^{\infty}f(t)(-p)e^{-pt}dt$$

$$= [e^{-p\infty}f(\infty) - e^{-p0}f(0)] + p\int_{0}^{\infty}f(t)e^{-pt}dt$$

$$= [0 - f(0)] + pF(p) = pF(p) - f(0)$$

(11) $\dfrac{df(t)}{dt}u(t) \Leftrightarrow pF(p) - f(0)$

Transform of an integral

if $f(t) \Leftrightarrow F(p)$ and $\int_0^t f(x)dx \Leftrightarrow F_1(p)$, then

$$F_1(p) = \mathcal{L}\left[\int_0^t f(x)dx\, u(t)\right] = \int_0^\infty u(t) \int_0^t f(x)dx\, e^{-pt}dt = \int_0^\infty \int_0^t f(x)dx\, e^{-pt}dt$$

if $u = \int_0^t f(x)dx$ and $dv = e^{-pt}dt$ then $du = f(t)dt$ and $v = -\frac{1}{p}e^{-pt}$

$$F_1(p) = \int_0^\infty udv = vu\Big|_0^\infty - \int_0^\infty vdu$$

$$F_1(p) = \left[-\frac{1}{p}e^{-pt}\int_0^t f(x)dx\right]_0^\infty - \int_0^\infty\left(-\frac{1}{p}\right)f(t)e^{-pt}dt$$

$$= \left(-\frac{1}{p}e^{-p\times\infty}\int_0^\infty f(x)dx\right) - \left(-\frac{1}{p}e^{-p\times 0}\int_0^0 f(x)dx\right) + \frac{1}{p}\int_0^\infty f(t)e^{-pt}dt$$

$$= \left(-\frac{1}{p}\times 0\int_0^\infty f(x)dx\right) - \left(-\frac{1}{p}\times 1 \times 0\right) + \frac{1}{p}F(p) = 0 - 0 + \frac{F(p)}{p}$$

(12) $u(t)\int_0^t f(x)dx \Leftrightarrow \dfrac{F(p)}{p}$

Transform of a time delay c (translation in time)

If $F(p) = \int_0^\infty f(t)u(t)e^{-pt}dt$, then let $F_c(p) = \int_0^\infty f(t-c)u(t-c)e^{-pt}dt$

let $\tau = t - c$

$$F_c(p) = \int_0^\infty f(\tau)u(\tau)e^{-p(\tau+c)}d\tau = e^{-pc}\int_0^\infty f(\tau)u(\tau)e^{-pt}d\tau = e^{-pc}F(p)$$

If $f(t)u(t) \Leftrightarrow F(p)$ then

(13) $f(t-c)u(t-c) \Leftrightarrow e^{-pc}F(p)$

Problem 2 Use the Laplace transform to solve for i(t) when initial conditions equal zero.

$$L\frac{di}{dt} + Ri + \frac{1}{C}\int_0^t i(t)dt = V_m \sin \omega t$$

A2.4 RL Circuit Transient Response

The Laplace transform operator is represented by the script letter $\mathscr{L}$. If the Laplace transform of f(t) is F(p) then F(p) is calculated from the integral transform (equation 2).

Figure 702 Series RL

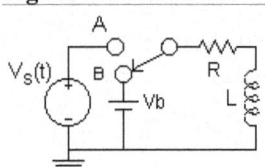

Define the Laplace Transform of any function f(t), from the time domain t to the complex frequency domain p, as a function F(p).

(2) $\quad F(p) = \mathscr{L}[f(t)] = \int\limits_0^\infty f(t)e^{-pt}\,dt$

The Laplace transform operator is implemented by multiplying each term in equation 14 (Figure 702) by exp(–pt), and integrating from 0 to ∞.

(14) $\quad v_S(t) = Ri(t) + L\dfrac{di}{dt} \qquad i(0) = \dfrac{V_b}{R}$

(15) $\quad e^{-pt}v_S(t) = e^{-pt}Ri(t) + e^{-pt}L\dfrac{di}{dt}$

(16) $\quad \int\limits_0^\infty e^{-pt}v_S(t)dt = \int\limits_0^\infty e^{-pt}Ri(t)dt + \int\limits_0^\infty e^{-pt}L\dfrac{di}{dt}dt$

We say the Laplace operator transforms $v_s(t)$ into $V_s(p)$, and i(t) into I(p).

(17) $\quad define\;\; V_s(p) = \int\limits_0^\infty e^{-pt}v_S(t)dt \qquad I(p) = \int\limits_0^\infty e^{-pt}i(t)dt$

Then the equation 14 transforms into equation 18 by applying the theorem for differentials (Section 7.3).

(18) $\quad V_S(p) = RI(p) + pLI(p) - Li(0)$

(19) $\quad V_S(p) = (R + pL)I(p) - Li(0)$

Now we can solve for I(p).

(20) $\quad I(p) = \dfrac{1}{(R+pL)}(V_S(p) + Li(0))$

Where i(0) is the current in L at time t=0 with the switch in position B.

Electronic Circuits – Practical Learning

Select a source We cannot proceed until we specify the $v_S(t)$ waveform, and transform it into $V_S(p)$. We arbitrarily select a sinewave as the forcing function. The source $v_S(t)$ is applied to the circuit at time t=0, because u(t) in effect moves the switch arm from position B to A at time t=0. Note how the exponential simplifies the math. Do not use trig functions sin and cos! (see superposition in 7.8)

$$\text{if } v_S(t) = V_M e^{j\omega t} u(t), \text{ then } V_S(p) = \frac{V_M}{p - j\omega} \text{ and}$$

$$(21) \quad I(p) = \frac{V_S(p)}{R + pL} + \frac{Li(0)}{R + pL} = \frac{1}{L} \cdot \frac{1}{p + \frac{R}{L}} \cdot \frac{V_M}{p - j\omega} + \frac{i(0)}{p + \frac{R}{L}}$$

Make a partial fraction expansion (Appendix A1) We need a partial fraction expansion, because the trick we use is based on the fact the inverse transform of 1/(p+a) is an exponential function (equation 4).

$$(4) \quad e^{-at} \Leftrightarrow \frac{1}{p + a}$$

$$(22) \quad I(p) = \frac{1}{L} \cdot \frac{1}{p + \frac{R}{L}} \cdot \frac{V_m}{p - j\omega} + \frac{i(0)}{p + \frac{R}{L}}$$

$$I(p) = \frac{1}{L} \cdot \frac{1}{j\omega + \frac{R}{L}} \cdot \frac{V_m}{p - j\omega} + \frac{1}{L} \cdot \frac{1}{p + \frac{R}{L}} \cdot \frac{V_m}{-\frac{R}{L} - j\omega} + \frac{i(0)}{p + \frac{R}{L}}$$

$$I(p) = \frac{V_m}{R + j\omega L} \cdot \frac{1}{p - j\omega} - \frac{V_m}{R + j\omega L} \cdot \frac{1}{p + \frac{R}{L}} + \frac{V_b}{R} \cdot \frac{1}{p + \frac{R}{L}}$$

Inverse transform A return from the *complex frequency domain p* to the time domain t is achieved by using the trick. The trick allows us to avoid calculating the inverse integral (equation 3). I(p) transforms to i(t) when we substitute an e^{-at} for each of the three 1/(p+a) terms. The complete solution consists of a steady-state term, a transient term due to the forcing function, and a transient term due to the natural response to the initial condition $i(0) = V_b/R$.

$$(23) \quad i(t) = \frac{V_m}{R + j\omega L} e^{j\omega t} - \frac{V_m}{R + j\omega L} e^{-\frac{R}{L}t} + \frac{V_b}{R} e^{-\frac{R}{L}t}$$

Time constant $\tau = L/R = 1/\omega_0$.

Laplace Transforms of vi Constraints

Time Domain:

component	R	L	C
voltage $v(t) =$	$Ri(t)$	$L\dfrac{di(t)}{dt}$	$\dfrac{1}{C}\displaystyle\int_0^t i(x)dx + v(0)$
current $i(t) =$	$\dfrac{1}{R}v(t)$	$\dfrac{1}{L}\displaystyle\int_0^t v(x)dx + i(0)$	$C\dfrac{dv(t)}{dt}$

Complex Frequency Domain: The responses i(t) or v(t) are not known. How do we know if the equations are Laplace transformable? *We do not know, however we proceed on the assumption they are Laplace transformable.*

Assume : $v(t) \Leftrightarrow V(p)$ $\qquad$ $i(t) \Leftrightarrow I(p)$

component	R	L	C
voltage $V(p) =$	$RI(p)$	$L[pI(p) - i(0)]$	$\dfrac{I(p)}{pC} + \dfrac{v(0)}{p}$
current $I(p) =$	$\dfrac{V(p)}{R}$	$\dfrac{V(p)}{pL} + \dfrac{i(0)}{p}$	$C[pV(p) - v(0)]$

Here they are again written as voltage and current transform pairs: f(t) $\Leftrightarrow$ F(p).

Voltage Transforms

$v(t)$	$\Leftrightarrow$	$V(p)$
$Ri(t)$	$\Leftrightarrow$	$RI(p)$
$L\dfrac{di(t)}{dt}$	$\Leftrightarrow$	$L[pI(p) - i(0)]$
$\dfrac{1}{C}\displaystyle\int_0^t i(x)dx + v(0)$	$\Leftrightarrow$	$\dfrac{I(p)}{pC} + \dfrac{v(0)}{p}$

Current Transforms

$i(t)$	$\Leftrightarrow$	$I(p)$
$\dfrac{v(t)}{R}$	$\Leftrightarrow$	$\dfrac{V(p)}{R}$
$\dfrac{1}{L}\displaystyle\int_0^t v(x)dx + i(0)$	$\Leftrightarrow$	$\dfrac{V(p)}{pL} + \dfrac{i(0)}{p}$
$C\dfrac{dv(t)}{dt}$	$\Leftrightarrow$	$C[pV(p) - v(0)]$

Electronic Circuits – Practical Learning

A3 Units

Dimensional analysis attaches a unit to every number or variable in an equation. Dimensional analysis is a skill all of us should have. This is a skill that enhances the ability to solve problems. In a way this claim is not to be believed until one uses dimensional analysis.

Unit Math The units attached to the numbers and variables are very important. One ascertains what units one wants in an answer and then works the problem backwards to figure out what one needs to solve it. This is done prior to doing anything with the numbers. Here is a very simple example.

You need to know how fast your car is moving in miles per hour (mph). when you know *it traveled one mile in one minute*. The first thing you need to do is ascertain the units of the answer. In this case it is miles per hour where per means "divided by".

$$answer = so\ many\ \frac{miles}{hour}$$

Transform the data, 1 mile per minute, into a format that will produce the units you want in the answer: Converting minutes into hours requires one knows that there are 60 min in an hour. In other words one has to know the conversion factors, which one can look up in tables. Units are variables, which can be cancelled according to the rules of algebra.

$$car\ speed = 1\frac{mile}{min} = 1\frac{mile}{min} \times 60\frac{min}{hour} = 60\frac{miles}{hour}$$

When all the units that can be cancelled are gone, one is left with 60 miles per hour, which is the correct answer.

Now, you might be saying to yourself that was easy. You are right! That is the point after all. If you follow this basic format, most of the "story problems" you encounter every day will be solved by dimensional analysis.

Here is another conversion problem. Convert the speed of light from meters per second to centimeters per nanosecond.

$$speed\ of\ light = 3\times10^8 \frac{meters}{sec} = 3\times10^8 \frac{meters}{sec} \times \frac{1}{10^9}\frac{sec}{ns} \times \frac{10^2}{1}\frac{cm}{meter} = 30\frac{cm}{sec}$$

Another application of this technique is solution verification. If the answer doesn't come out in the right units, most likely something was wrong in your calculation. Always put units on the numbers and equations you use. That way when you see the correct units at the end of your work, it confirms that the equations are set up properly. So, whenever you come upon a question that seems to have a whole pile of data and you have no idea where to begin, first figure out which units you want the answer in. Then shape that pile of data until the units match the units needed for the answer.

Rules of Thumb
■ Always add units to your equations. They give you the means to be confident you have the right answer.

■ Use units to create an equation that solves the problem. Do this by adding units to the numbers and variables, and then canceling units until the result appears.

■ Use estimation to determine approximately what the answer should be as you are analyzing. Then compare the estimate to the results to identify mistakes.

Another way to solve differential equations is to assume a solution.

assume the solution to the equation is $i(t) = I_0 e^{-at}$

$$v_m = RI_0 e^{-at} - \frac{1}{aC} I_0 e^{-at} \Rightarrow v_1(0^-) = 0 = \left(R - \frac{1}{aC}\right) I_0 \Rightarrow a = \frac{1}{RC}$$

then $v_m = RI_0 e^{-\frac{t}{RC}} + \frac{1}{C} \int_0^t i(x)dx = RI_0 e^{-\frac{t}{RC}} + I_0 \left(-\frac{RC}{C}\right) e^{-\frac{x}{RC}} \Big|_0^t$

$$v_m = I_0 \, \text{Re}^{-\frac{t}{RC}} - I_0 \, \text{Re}^{-\frac{t}{RC}} + I_0 R = I_0 R$$

$$\therefore I_0 = \frac{v_m}{R} \Rightarrow v_m = \frac{v_m}{R} \text{Re}^{-\frac{t}{RC}} - \frac{v_m}{R} \text{Re}^{-\frac{t}{RC}} + \frac{v_m}{R} R = v_m \quad qed$$

and $i(t) = \frac{v_m}{R} e^{-at}$ *is a solution*

also $v_1(t) = v_R + v_C \Rightarrow v_C = -\frac{v_m}{R} \text{Re}^{-\frac{t}{RC}} + \frac{v_m}{R} R = v_m \left(1 - e^{-\frac{t}{RC}}\right)$ *Figure 70411*

Index